T0296535

CAMBRIDGE STUDIES IN
ADVANCED MATHEMATICS: 51

SPINNING TOPS

Already published

1 W.M.L. Holcombe *Algebraic automata theory*
2 K. Petersen *Ergodic theory*
3 P.T. Johnstone *Stone spaces*
4 W.H. Schikhof *Ultrametric calculus*
5 J.-P. Kahane *Some random series of functions, 2nd edition*
6 H. Cohn *Introduction to the construction of class fields*
7 J. Lambek & P.J. Scott *Introduction to higher-order categorical logic*
8 H. Matsumura *Commutative ring theory*
9 C.B. Thomas *Characteristic classes and the cohomology of finite groups*
10 M. Aschbacher *Finite group theory*
11 J.L. Alperin *Local representation theory*
12 P. Koosis *The logarithmic integral I*
13 A. Pietsch *Eigenvalues and s-numbers*
14 S.J. Patterson *An introduction to the theory of the Riemann zeta-function*
15 H.J. Baues *Algebraic homotopy*
16 V.S. Varadarajan *Introduction to harmonic analysis on semisimple Lie groups*
17 W. Dicks & M. Dunwoody *Groups acting on graphs*
18 L.J. Corwin & F.P. Greenleaf *Representations of nilpotent Lie groups and their applications*
19 R. Fritsch & R. Piccinini *Cellular structures in topology*
20 H Klingen *Introductory lectures on Siegel modular forms*
22 M.J. Collins *Representations and characters of finite groups*
24 H. Kunita *Stochastic flows and stochastic differential equations*
25 P. Wojtaszczyk *Banach spaces for analysts*
26 J.E. Gilbert & M.A.M. Murray *Clifford algebras and Dirac operators in harmonic analysis*
27 A. Frohlich & M.J. Taylor *Algebraic number theory*
28 K. Goebel & W.A. Kirk *Topics in metric fixed point theory*
29 J.E. Humphreys *Reflection groups and Coxeter groups*
30 D.J. Benson *Representations and cohomology I*
31 D.J. Benson *Representations and cohomology II*
32 C. Allday & V. Puppe *Cohomological methods in transformation groups*
33 C. Soulé et al *Lectures on Arakelov geometry*
34 A. Ambrosetti & G. Prodi *A primer of nonlinear analysis*
35 J. Palis & F. Takens *Hyperbolicity, stability and chaos at homoclinic bifurcations*
37 Y. Meyer *Wavelets and operators I*
38 C. Weibel *An introduction to homological algebra*
39 W. Bruns & J. Herzog *Cohen-Macaulay rings*
40 V. Snaith *Explicit Brauer induction*
41 G. Laumon *Cohomology of Drinfield modular varieties I*
42 E.B. Davies *Spectral theory and differential operators*
43 J. Diestel, H. Jarchow & A. Tonge *Absolutely summing operators*
44 P. Mattila *Geometry of sets and measures in Euclidean spaces*
45 R. Pinsky *Positive harmonic functions and diffusion*
46 G. Tenenbaum *Introduction to analytic and probabilistic number theory*
47 C. Peskine *An algebraic introduction to complex projective geometry I*
48 Y. Meyer & R. Coifman *Wavelets and operators II*
49 R. Stanley *Enumerative combinatories*
50 I. Porteous *Clifford algebras and the classical groups*
51 M. Audin *Spinning tops*
52 V. Jurdjevic *Geometric control theory*
53 H. Voelklein *Groups as Galois groups*
54 J. Le Potier *Lectures on vector bundles*
55 D. Bump *Automorphic forms*
56 G. Laumon *Cohomology of Drinfeld modular varieties II*
57 D. Clarke & B.A. Davey *Natural dualities for the working algebraist*
59 P. Taylor *Practical foundations of mathematics*
60 M. Brodmann & R. Sharp *Local cohomology*
64 J. Jost & X. Li-Jost *Calculus of variations*

Spinning tops

A Course on Integrable Systems

Michèle Audin

Institute de Recherche Mathématique Avancée
Université Louis Pasteur et CNRS

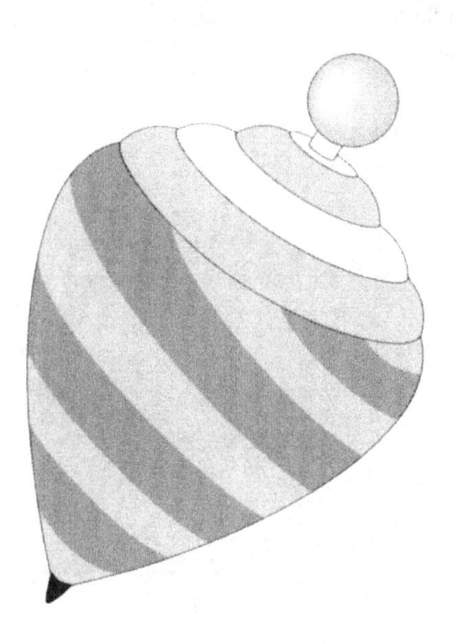

PUBLISHED BY THE PRESS SYNDICATE OF THE UNIVERSITY OF CAMBRIDGE
The Pitt Building, Trumpington Street, Cambridge, United Kingdom

CAMBRIDGE UNIVERSITY PRESS
The Edinburgh Building, Cambridge CB2 2RU, UK http://www.cup.cam.ac.uk
40 West 20th Street, New York, NY 10011–4211, USA http://www.cup.org
10 Stamford Road, Oakleigh, Melbourne 3166, Australia
Ruiz de Alarcón 13, 28014 Madrid, Spain

First published 1996
First paperback edition 1999

Typeface Computer Modern 12/13pt *System* LaTeX [UPH]

A catalogue record for this book is available from the British Library

ISBN 0 521 56129 9 hardback
ISBN 0 521 77919 7 paperback

Transferred to digital printing 2004

Contents

Acknowledgements

Among the works in which I have learned the most, there are three papers that I wish to mention here. I think that I was very lucky to start my initiation to integrable systems by reading two very beautiful papers that are not among the most cited, those of Verdier [84] and Griffiths [36]. On the one hand, Verdier had the brilliant idea of illustrating his exposition of the work of Adler and van Moerbeke by the example of the symmetric top. On the other hand, Griffiths' main philosophical point was to look at a Lax equation, without specifying anything more. The advantage of this approach is that it allows one to start working without having first ingested loop algebras and the "AKS theorem". Moreover, in this sober presentation, the role played by the eigenvectors of the Lax matrices is amply brought to light. The algebraic geometry related to these eigenvectors is extremely well described in Reyman's paper [74] – which is the third paper to which I feel indebted.

The present text originates mainly from several talks[1] I have given on the examples here, in particular on the work I have done jointly with R. Silhol [15] and from a graduate course I taught in Strasbourg in 1992-93, jointly with J.-Y. Mérindol, on "Algebraic curves and integrable systems". The first version, *Toupies, un cours sur les systèmes intégrables* was written at the end of 1993. I am very pleased to acknowledge the influence of the very clear survey of Reyman & Semenov-Tian-Shanski [77] on the present version.

I have learned a lot in discussions with Jean-Yves Mérindol, Robert Silhol, and, especially, Alexei Reyman.

A lot of people have helped me to understand the material here, by criticism, questions or simply remarks during a talk, among which are, colleagues or students, Nicole Bopp, Nicole Desolneux-Moulis, Ljubomir Gavrilov, Sophie Gérardy, Bertrand Haas, Patrick Iglesias, Viatcheslav Kharlamov, Dimitri Markushevich, Nguyen Tien Dũng, Nitin Nitsure, Leonid Polterovich, Claude Sabbah, Jean-Marie Strelczyn, Jean Stutzmann and Pol Vanhaecke. I wish to thank them all.

The pictures in this book have been created by Raymond Seroul, whom I am also very pleased to thank.

Michèle Audin
Strasbourg, January 30, 1996

---

[1] I thank all the people who have invited me to give or listened to me giving talks in Basel, Bochum, Bombay, Boston, Cambridge, Haifa, Lausanne, Luminy, Lyon, Montréal, Paris, Nantes, Strasbourg, Tel-Aviv.

Last, but not least, I wish to thank the staff of Cambridge University Press, especially David Tranah, who kindly welcomed the book, and Susan Parkinson, who was very helpful in improving the English.

Michèle Audin
Strasbourg, January 8, 1996

For the paperback edition, I have only corrected a few misprints.

Strasbourg, May 1, 1999

Introduction

«On donne deux circonférences O et O'. D'un point A pris sur O on mène les tangentes à O'; on joint les points de contact de ces tangentes; on mène la tangente en A à la circonférence O. On demande le lieu du point d'intersection de cette tangente avec la corde des contacts de la circonférence O'.»
Chacun comprenait l'importance d'un pareil théorème.

JULES VERNE
Paris au vingtième siècle

The objective of this book is to give an idea of some modern techniques in the theory of integrable systems, and of how to use them to get topological information, through examples coming from mechanics. I have tried to go straight to the point, explaining the techniques as I go along and delaying the general theory to the appendices.

Our main topic is a "rigid-body-with-a-fixed-point-in-a-constant-gravitational-field", in the cases where the differential equations describing its motion are completely integrable. It is a recent custom – without doubt both unfortunate and convenient – to abbreviate this cumbersome locution to *spinning top* or even *top*. I will follow this.

Spinning tops are celebrated examples of completely integrable systems with two degrees of freedom: a rigid body moving about its centre of mass, a rigid body with an axis of revolution (this is the case that explains – rather badly – the use of the words "spinning top") and lastly the mysterious (?) Kowalevski case.

Much work has been done during the two last decades on integrable systems in both finite and infinite dimensions and a lot of sophisticated techniques have been developed, so that these systems henceforth stand at a crossroads of numerous avenues of mathematics. I will only mention here representation theory (of Lie algebras, loop algebras, Kac-Moody algebras ...) and algebraic geometry (algebraic curves, Abelian varieties, ϑ-functions ...).

In the meantime, but quite separately, some studies of the topology of these systems (Liouville tori and their bifurcations) have appeared.

My starting point was the idea that the "sophisticated techniques" I mentioned might be powerful enough to give some information on the topology as well. This is actually the case and this is what I aim to show in this text. Here the originality is primarily in the approach, in the common method used to handle the different cases. However, as this method is, after all, rather natural, it eventually gives new results about old problems ...

Spinning tops have been investigated since the eighteenth century and it is well known that the solutions of the differential equations can be expressed in terms of elliptic functions or, in the Kowalevski case, of Abelian functions related to a hyperelliptic curve.

Thus, a posteriori, it is known that there are algebraic curves in the landscape. Modern techniques bring them to the fore a priori: there is indeed an algebraic curve, we know it from the beginning; it can be a way to express that the system has enough constants of motions, first integrals; and it can be used to write down the solutions. What I will explain here is how it can also be used to describe the topology of the system.

As an introduction, I will try to give now a rough idea of the method, in a rather discursive style – some of the missing details will be developed in the appendices.

1. Completely integrable systems

A completely integrable system is a Hamiltonian system that admits the maximum possible number of first integrals. The easiest way to be more precise is to state a definition in a symplectic manifold, but it turns out that the natural scenery of almost all examples is a Poisson manifold (a very careful exposition of the Poisson case can be found in Vanhaecke's paper [83]). I will recall here some basic definitions and results, and will content myself with giving hints and references for some of the notions and proofs.

1.1. Poisson structures and Hamiltonian systems

A Poisson manifold is a smooth manifold W, the ring of functions $C^\infty(W)$ of which is endowed with a Lie algebra structure $\{\,,\,\}$ which is a derivation in both its entries. That is, $\{\,,\,\}$ must be skew-symmetric, satisfy the Jacobi identity

$$\{f,\{g,h\}\}+\{g,\{f,h\}\}+\{h,\{f,g\}\}=0$$

and the Leibniz rule

$$\{f,gh\}=\{f,g\}\,h+g\,\{f,h\}\,.$$

The bracket $\{\,,\,\}$ is the *Poisson bracket*. As any function H on W defines a derivation $\{H,\cdot\}$, it defines a vector field X_H, *the Hamiltonian vector field:* by definition

$$X_H\cdot f=\{H,f\}=df\,(X_H)\,.$$

For instance, in the case of spinning tops, W is the vector space $\mathbf{R}^3\times\mathbf{R}^3$, which we shall consider as "the dual of the Lie algebra of the group of Euclidean motions of the 3-space". This seems to be very intricate, but gives it a Poisson-manifold natural structure: the duals of the Lie algebras are the basic Poisson manifolds (see Appendix 1).

The function H (the Hamiltonian) defines a vector field X_H (the Hamiltonian vector field), which in turn defines a differential equation, the *Hamiltonian system* associated with H.

By skew-symmetry, $\{H,H\}=0$ so that $dH(X_H)=0$ and H is constant along the trajectories of X_H. In other words, the solutions of the Hamiltonian system remain in the levels of H. In "Hamiltonian" mechanics, when H is the total energy of a system, H is constant during the motion.

Any function f having that property (remaining constant along the trajectories), that is, such that $df(X_H)=0$ (or $\{H,f\}=0$) is called a first integral. Notice that a Hamiltonian system always has at least one first integral!

For instance, if $f\in C^\infty(W)$ is such that $\{f,g\}=0$ for any function g on W, it is a first integral for *any* Hamiltonian system on W. Such functions are called *Casimir functions* – or "trivial" first integrals in the physicists' terminology.

1.2. Completely integrable systems on symplectic manifolds

An important subclass of Poisson manifolds is that of *symplectic* manifolds. A symplectic manifold is, roughly speaking, a manifold W endowed with a non-degenerate 2-form ω. Being non-degenerate, ω allows us to associate, with any function H, a vector field X_H defined by

$$\omega\left(X_H, \cdot\right) = dH\left(\cdot\right).$$

This defines in turn a bracket on $\mathcal{C}^\infty(V)$:

$$\{f, g\} = \omega\left(X_g, X_f\right) = dg\left(X_f\right)$$

which is obviously skew-symmetric and satisfies the Leibniz rule. An exercise that is both easy and useful is to check that this bracket satisfies the Jacobi identity if and only if the form ω is *closed*, and this is the reason why a symplectic manifold is a manifold endowed with a non-degenerate closed 2-form.

Of course, the non-degeneracy condition forces the dimension of the manifold to be even, and this is enough to prove that there are much more Poisson than symplectic manifolds (there is a canonical non-trivial Poisson structure on, say, $\mathbf{R}^3$; see the examples in Appendix 1). Nevertheless, symplectic manifolds are Poisson manifolds, functions on them define Hamiltonian systems[1] and we come back to first integrals.

DEFINITION. *A Hamiltonian system on a $2n$-dimensional symplectic manifold is completely integrable (or simply integrable) if it has n functionally independent first integrals $H_1, \ldots, H_n$ that pairwise commute (i.e. such that $\{H_i, H_j\} = 0$).*

Remarks. Write $X_i = X_{H_i}$.

- "Functionally independent" means that there exists an open dense subset U in the manifold V such that, for $x \in U$, the differentials $dH_i(x)$ (or, which amounts to the same thing, the vectors $X_i(x)$) are independent.

- Since H_i and H_j commute, $\omega(X_i, X_j) = 0$: at a generic point of V, the X_i will generate an n-dimensional isotropic subspace of the tangent space T_xV. Since $\dim V = 2n$, n is the maximum possible dimension for isotropic subspaces, so that n is also the maximum possible number of (independent) first integrals. For instance, the Hamiltonian of the system belongs to the algebra generated by the functions H_i.

- A classical exercise (see any textbook on symplectic geometry, e.g. that of Libermann & Marle [59]) is to show that $[X_f, X_g] = \pm X_{\{f,g\}}$, in other words that the vector fields X_i commute.

- Last but not least, I do insist on the fact that *any* function in the algebra generated by $H_1, \ldots, H_n$ will give a Hamiltonian system that admits $H_1, \ldots, H_n$ as first integrals and thus will be completely integrable. There is no reason at this point to require that H_1 should be the Hamiltonian of the system under consideration. What is

[1] In the symplectic manifold $\mathbf{R}^{2n}$ with coordinates $(p_1, \ldots, p_n, q_1, \ldots, q_n)$ and symplectic form $\sum dp_i \wedge dq_i$, the equations in the Hamiltonian system associated with H are the celebrated "Hamilton equations". Hence all the terminology used here.

intrinsically defined is actually the algebra generated by the functions H_i, but it will be convenient to use a basis. Once a basis is chosen, we may consider it as a *momentum mapping*

$$H = (H_1, \ldots, H_n) ; W \longrightarrow \mathbf{R}^n$$

since this is the momentum mapping for the local $\mathbf{R}^n$-action defined by the flows of the functions H_i.

1.3. A few words on integrable systems on Poisson manifolds

A Poisson manifold is foliated by submanifolds on which the Poisson bracket defines a Poisson structure (it should be noticed that the Poisson bracket defines nothing on a submanifold in general), which is associated with a symplectic form: in short, *symplectic leaves.* I do not want to spend much time on the details of the formulation or the proof[2] of this result.

The simplest case is when the symplectic foliation is described by the Casimir functions. This happens quite often, in particular in the examples we have in view. More precisely, this is the case when the generic leaves are the connected components of the regular common levels of Casimir functions. Then, if $2n$ is the maximal dimension[3] of symplectic leaves and if $2n + m$ is the dimension of the Poisson manifold, the maximum number of independent commuting first integrals one can expect is $m + n$ (m "trivial", n "non-trivial"). Moreover, as the Casimirs commute with everything, the non-trivial integrals define commuting flows on the symplectic leaves in this case. The cases we will consider are those involving orbit foliation for the coadjoint action on the dual of a Lie algebra, for Lie algebras in which that property is satisfied. This is the subject of Appendix 1.

In any case, we can restrict ourselves to symplectic leaves, so that we can content ourselves with the "symplectic" definition above.

Remark. The consideration of general Poisson structures, and even of several Poisson structures on the same manifold, is very useful in *constructing* integrable systems (see e.g. the discussion about the AKS theorem in Appendix 2).

1.4. A list of examples

Finite-dimensional integrable systems include the following[4] examples: the Calogero-Moser systems, the Calogero-Sutherland systems, the Calogero systems, the Clebsch rigid body in an ideal fluid, the dimension-n rigid body, the Euler-Arnold rigid body, the Euler equations, the Euler-Poinsot top, the exotic $SO(4)$-top, the free particle on an ellipsoid, the free rigid body, the Garnier system, the Gaudin system, the geodesic flow on a torus, the geodesics on an ellipsoid, the geodesics on a surface of revolution, the geodesics on quadrics, the geodesics on $SO(3)$, the Goldman functions, the Goryachev-Chaplygin top, the harmonic oscillator, the Hénon-Heiles system, the Holt potential, the

[2] I suggest that the interested reader should look for instance at Libermann & Marle [59] and at the references therein.

[3] Of course, the foliation is singular.

[4] I have used in particular Reyman & Semenov-Tian-Shanski's survey [77], Perelomov's book [70], de Dinteville's paper [25] and my lecture notes [14] to compile this list, which is both incomplete and redundant.

Jeffrey-Weitsman system, the Kepler problem, the Kirchhoff rigid body in an ideal fluid, the Kolosoff potential, the Kovalevskaya[5] rigid body, the Kowalevskaya gyroscope, the Kowalevski top, the Lagrange top, the mathematical pendulum, the Moser systems, the motion of a particle in a central field, the motion of a particle in a potential field, the motion of a particle on a sphere in a quadratic potential, the Neumann problem, the nonabelian Toda lattices, the non-periodic Toda lattice, the partially symmetric gyroscope, the pendulum, the periodic Toda lattice, the Ruijsenaars system, the spherical pendulum, the $SO(n)$ top, the Steklov rigid body in an ideal fluid, the symmetric top, the Toda lattice, the two-body problem, the two-dimensional anharmonic oscillator, the two-dimensional oscillator.

2. The Arnold-Liouville theorem

2.1. What the Arnold-Liouville theorem says

Now we are back in a symplectic manifold V and we have an integrable system as in the definition above. Thus the geometric situation is the following: at a point x of the open dense subset where the integrals H_i are independent, we have n independent vectors; they span a Lagrangian[6] subspace of T_xV that is tangent to the common level of $H_1, \ldots, H_n$ in which x stands. If there is any reason of thinking that the flows of the vector fields are complete, we may derive an action of the group $\mathbf{R}^n$ (recall that the vector fields do commute), on any regular level, that is locally free (independence again). Now this gives to the regular levels an affine structure in which the flows under consideration are linear. This is what the Arnold-Liouville theorem says.

THEOREM (Arnold [9]). *Let $h = (h_1, \ldots, h_n) \in \mathbf{R}^n$ be a regular value of the mapping $(H_1, \ldots, H_n) : V \to \mathbf{R}^n$. Let $\mathcal{T}_h$ be the corresponding regular level, so that $\mathcal{T}_h$ is a Lagrangian submanifold. Let x be a point in $\mathcal{T}_h$. If the flows of the vector fields $X_1, \ldots, X_n$ starting at x are complete, the connected component of x in $\mathcal{T}_h$ is a homogeneous space of $\mathbf{R}^n$. In particular, it has coordinates $(\varphi_1, \ldots, \varphi_n)$ in which the vector field X_i can be written:*

$$X_i = \sum_{j=1}^{n} \omega_j^i(h) \frac{\partial}{\partial \varphi_j}.$$

Remark. This is actually the easy half of the Arnold-Liouville theorem. The most economical way to ensure that the flows are complete is to require that the component of x is compact (this will be so if one of the integrals is proper). Of course, in this case, it must be a torus (one of the celebrated Liouville tori) and $(\varphi_1, \ldots, \varphi_n)$ can be considered as mod 2π coordinates. This is why they are called *angle* coordinates. Now the Arnold-Liouville theorem states that there are some complementary coordinates, *action*[7] coordinates: they are related to another affine structure, transversal to the levels. I will not discuss action coordinates here, despite their great significance.

[5] As a general rule, I have tried to use the authors' names as they appear in their papers published in French or English, e.g. Sophie Kowalevski.

[6] i.e. isotropic for the symplectic form, and of maximal dimension.

[7] The existence of the action-angle coordinates may be understood as a precise way of saying that the differential system is "integrable by quadratures".

2.2. What the Arnold-Liouville theorem does not say

From the topological viewpoint, the theory seems to be rather poor: there are only tori, and moreover the solutions of the differential systems are (images of) straight lines. Nevertheless some non-trivial questions remain.

1. How is it possible to decide, conceptually (i.e. without too much computation) that an actual level h is regular? In the simple case of spinning tops, as we shall see, the discussion is about a common level of four polynomials in six variables ....

2. How will we know whether the flows are complete?

3. And even is a level is known to be compact, of how many Liouville tori does it consist?

4. What happens if one goes through a critical value?

5. It is quite tautological to claim that the flows are linear with respect to an affine structure that they have themselves defined. Is there any linearisation statement with respect to a more canonical affine structure?

I will now sketch the method that I want to discuss and illustrate in this text. Its main feature is that it provides a framework in which there are natural strategies for investigating and even for answering all the above questions. The idea is to model the levels together with their non-explicit affine structure by objects that are rigid enough to be endowed with a *canonical* affine structure. These are the Abelian varieties. This is no surprise: even if we did not notice them, they appeared in the landscape when I mentioned elliptic functions and/or Abelian integrals.

3. A discourse on the method

Forget integrable systems for the moment. The method will apply to *Lax equations*, that is, differential equations of the form

$$\frac{d}{dt}A = [A, B]$$

where A and B are real or complex matrices depending on time. The bracket is the usual Lie bracket of matrices, so that such an equation expresses at the infinitesimal level the fact that the matrix A remains in the same conjugacy class, in other words that the solutions have the form

$$A(t) = U(t)A(0)U(t)^{-1}$$

for some (unknown!) invertible matrix $U(t)$.

Remark. Notice that, in general, the unknown functions that are the entries of A also appear in the entries of B. In other words, B will be a function of A: despite the notation, the differential equation is nonlinear.

The method I am discussing will actually apply to Lax equations in which A and B have entries in the ring of real or complex Laurent polynomials in a variable λ, which will be called the *spectral parameter*. I prefer to put the dependence on λ into the notation, so that A and B will be called A_λ and B_λ and the equation will have the form

$$\frac{d}{dt}A_\lambda = [A_\lambda, B_\lambda],$$

where the bracket is again that of matrices: $[A_\lambda, B_\lambda] = A_\lambda B_\lambda - B_\lambda A_\lambda$.

For instance, the differential equation

$$\frac{d}{dt}\begin{pmatrix} b_1 & 1 & x_3\lambda \\ x_1 & b_2 & 1 \\ \lambda^{-1} & x_2 & b_3 \end{pmatrix} = \left[\begin{pmatrix} b_1 & 1 & x_3\lambda \\ x_1 & b_2 & 1 \\ \lambda^{-1} & x_2 & b_3 \end{pmatrix}, \begin{pmatrix} 0 & 0 & x_3\lambda \\ x_1 & 0 & 0 \\ 0 & x_2 & 0 \end{pmatrix}\right]$$

is a system of twenty-seven equations (coefficients of λ^{-1}, 1, λ in each entry) in six unknown functions, which turns out to be equivalent to

$$\begin{cases} \dot{x}_1 &= x_1(b_2 - b_1) \\ \dot{x}_2 &= x_2(b_3 - b_2) \\ \dot{x}_3 &= x_3(b_1 - b_3) \end{cases} \qquad \begin{cases} \dot{b}_1 &= x_1 - x_3 \\ \dot{b}_2 &= x_2 - x_1 \\ \dot{b}_3 &= x_3 - x_2 \end{cases}$$

3.1. The algebro-geometric zoo

One has the right to ask the reasons for doing something so intricate (having added the spectral parameter for instance), and also how to put a given differential equation into this form. I will not discuss the last question here but, rather, I will explain what the Lax form and the spectral parameter give us.

A considerable number of first integrals. Any Lax equation has, a priori, many first integrals: since the matrix A stays in the same conjugacy class, its eigenvalues will be "constants of motion". In other words, the coefficients of the characteristic polynomial of A are first integrals.

A curve. Now, if there is a spectral parameter in the matrix A, the characteristic polynomial is a polynomial in *two* variables

$$P(\lambda, \mu) = \det(A_\lambda - \mu\,\mathrm{Id}).$$

But what is a polynomial in two variables if not the equation of an algebraic curve? The complex curve C of equation

$$P(\lambda, \mu) = 0$$

will be called the *spectral curve:* it describes the eigenvalues, the *spectrum* of the matrices A_λ. The coefficients in the equation of C are the first integrals mentioned above, so that there are actually *many* spectral curves: one for each value of the set of integrals (in other words, the equation $P(\lambda, \mu) = 0$ describes a family of curves). One could or should index the curve by the value to which it corresponds; once we have noticed that a spectral curve corresponds to a common level of the first integrals, we can also index a level by a curve: I will sometimes write $\mathcal{T}_C$ for the level (in the set of matrices) corresponding to the curve C.

The leading figure. Fix a curve C in the family and suppose that, for some value λ of the spectral parameter, A_λ has a simple spectrum. For any eigenvalue μ of A_λ (that is to say, any μ such that $(\lambda, \mu) \in C$) we now have a line in the space $\mathbf{C}^N$ on which our matrices act: the one-dimensional eigenspace of A_λ with respect to μ. Let us now allow λ to vary. Putting all these lines together, what we get is a complex line bundle on the curve C, the *eigenvector bundle*, the fibre of which at a point (λ, μ) is the eigenspace[8] of A_λ with respect to μ.

Now, when A_λ varies in the common level $\mathcal{T}_C$, we can consider all these line bundles together as a mapping

$$\varphi = \varphi_C : \mathcal{T}_C \longrightarrow \operatorname{Pic}(C)$$

with values in the Picard group of (algebraic) complex line bundles over C (see Appendix 4). This map, the *eigenvector mapping*, will be both our main tool and the leading figure in this text.

Remark. Recall now that any component of the Picard group is a complex torus, so that it shows up in the scenery with its own affine structure (which, incidentally, is canonical). One can dream of reading possible angle variables as the linearisation of flows in $\operatorname{Pic}(C)$. More precisely, if $t \mapsto A_\lambda(t)$ is a solution of the Lax equation, let us consider the image curves $t \mapsto \varphi(A_\lambda(t))$. To state that they are straight lines, with linear parametrisation, would not be tautological, the affine structure underlying the statement being defined without any reference to the vector field in question.

I must also emphasise that the first integrals, the curve and the eigenvector mapping itself do not depend on the matrix B_λ. This is no surprise: in the case of an integrable system $(H_1, \ldots, H_n)$, to each function in the algebra generated by $(H_1, \ldots, H_n)$ will correspond a different B_λ. Our algebraic data describe the levels as a whole, not a special flow on them.

3.2. The case of an integrable system

All this discussion applies very readily to any Lax equation, but we want to apply it to an integrable system. Of course, though Lax equations always have a lot of first integrals, nothing forces these to commute (we do not even have a Poisson structure to give a meaning to this), and, of course, there might not be enough of them. Actually, there is a machinery that allows us to construct Lax equations that are integrable systems: this is the so-called Adler-Kostant-Symes theorem (see Appendix 2). For the moment, I will just *assume* that the Lax equation under consideration is indeed an integrable system. Also, to simplify the discussion, I will often assume that the flows are complete. In this case, the level $\mathcal{T}_C$ is a homogeneous[9] space of some $\mathbf{R}^n$, which is, moreover, described by polynomial equations. The complexified level $\mathcal{T}_C^{\mathbf{C}}$ (the complex solutions of the same polynomial equations) could thus perfectly well be an open subset of a complex torus.

---

[8]There always exist special values of λ for which the spectrum of A_λ is not simple. However the complex line bundle is well defined, at least if the curve C is smooth (see Appendix 3).

[9]I am implicitly assuming that the matrices A_λ and B_λ are real matrices, i.e. have real entries, although I make them act on the complex vector space $\mathbf{C}^N$.

To dream. Can we use the eigenvector mapping φ_C to answer all the questions raised in 2.2? If the differential system were real, the mapping φ_C would preserve all real structures and send $\mathcal{T}_C$ (the real part of $\mathcal{T}_C^{\mathbf{C}}$) into the real part of Pic(C), which is quite easy to describe.

To wake up. The eigenvector mapping φ_C can almost never be an isomorphism.

1. Dimensions may not fit. First of all, there is no reason why $\mathcal{T}_C^{\mathbf{C}}$ and Pic(C) should have the same dimension. In the case of the rigid body, $\dim \mathcal{T}_C = 2$, but we shall get Picard groups of dimension 1 for the symmetric (Lagrange) top and 5 for the Kowalevski top. Notice that $\dim \mathcal{T}_C$ is a datum that comes with the integrable system: it is half the dimension of the symplectic (phase) manifold – what physicists call the number of degrees of freedom. On the contrary, $\dim \operatorname{Pic}(C)$ is the genus of the spectral curve (see Appendix 4) and is related to both the degree in λ of A_λ and the size N of the matrices: it depends on the Lax form we have used. However, it can happen that the dimensions are the same: it is the case for systems such as geodesics on an ellipsoid, the Neumann problem[10] and the periodic Toda lattice, as also for the "dimension-3 free rigid body", a one-degree-of-freedom system, the first to be investigated in this book.

2. The compactness problem. The mapping φ_C cannot be a complex isomorphism, simply because $\mathcal{T}_C^{\mathbf{C}}$ is defined by polynomial equations in a linear space and is an *affine* algebraic variety while the components of Pic(C) are (compact) tori. Even if the dimensions fitted, φ_C could, at best, send $\mathcal{T}_C^{\mathbf{C}}$ onto an open subset of Pic(C).

3. The real part. Let us assume that one of the first integrals is proper, so that the real level $\mathcal{T}_C$ is compact. In general, the real part of Pic(C) will intersect the above mentioned hypersurfaces, so that even the restriction of φ_C to the real part cannot be an isomorphism onto its image.

What happens quite often is that $\dim \mathcal{T}_C \leq \dim \operatorname{Pic}(C)$ and φ_C is a finite covering of an open subset in an Abelian subvariety of Pic(C). A very common situation for two degrees of freedom is that of a genus-3 spectral curve endowed with an involution τ with four fixed points and where, for some reason, it is known that φ_C takes its values in the "anti-fixed" points of τ, the Abelian variety Prym(τ) (see Appendix 5).

3.3. What can nevertheless be done

Regular levels and linearisation. Questions 1 and 5 that I have raised in 2.2 have natural answers in this framework. It is possible to discuss in great generality (that is, for a Lax equation, with no additional integrability assumption) whether φ_C linearises the flow. In almost all the cases we will need to consider, the matrix B_λ occurring in the Lax equation will have a very specific form, so that the eigenvector mapping will automatically linearise the solutions of the Lax equation (see the work of Griffiths [36] and Reyman [74] as well as Appendix 3). These linearisation theorems are simple consequences of determining the tangent mapping of φ_C, according to which it is very often possible to prove statements like "if C is smooth, the corresponding level is regular". Notice that nothing prevents the spectral curve from being singular for all values of the first integrals (a very natural example of this situation will be explained in III.3.2).

[10] See the papers of Moser [63], Knörrer [52] and the author [11, 13].

Liouville tori and their bifurcations. Assume now that one of the first integrals is proper (this is the case for the total energy in the case of tops), so that the levels are compact and the flows complete. Look at Questions 3 and 4 of 2.2. Of course, the fact that φ_C is a covering of its image forbids its direct use to enumerate the Liouville tori. However, it is often possible to modify it, or to understand it well enough to use it to identify the levels with the real part of an Abelian variety – but one has to find another trick for any other case. I have already mentioned that the real part of a Jacobian is easy to investigate once the real structure of the curve itself is well understood (see Appendix 4). The case of a Prym is somewhat more difficult but is still quite feasible.

Varying the values of the first integrals, one gets *families* of Abelian varieties, which may be used to study the critical levels and the bifurcations of the Liouville tori.

Non-compactness. In the general case of a Lax equation, the levels $\mathcal{T}_C$ are common levels of a family of polynomials defined on a linear space, so that the complexified levels are never compact, as I have already mentioned. The image of the eigenvector mapping in the (complex, compact) Jacobian is in general well understood: very often, the complement is related to the Θ-divisor of the curve (see Appendix 4). To say that the (real) levels are non-compact amounts to saying that their image meets the "divisor at infinity": we thus still have the technology to describe the topology in this case (see Chapter V).

3.4. Another approach

There is another possible approach, which turns out to be dual to the one discussed so far: one can consider directly a (complex) level set as an affine algebraic variety, "simply" by adding a divisor at infinity and proving that the completed variety is an Abelian variety, on which the Hamiltonian vector field can be extended in a constant (that is, linear) vector field. In general, the method used to obtain this divisor at infinity is to look for all the Laurent series (in time) that are formal solutions of the system, just by substitution; then the divisor is the locus of the poles. This method relies on algebraic computations that are often tedious. Moreover, the determination of these points at infinity uses a specific vector field to depict the common level sets of several functions, which I find quite unsatisfying. In its favour, the method shows the level set directly in an Abelian variety, avoiding the problem of coverings (there are many papers on this subject, e.g. those of Haine [39], Adler & van Moerbeke [5, 6] and Vanhaecke [82]).

3.5. And if there is no spectral parameter?

All the constructions above make an essential use of the spectral parameter λ: the latter is responsible for the existence of the spectral curve and hence of all the algebro-geometric zoo presented here. There are honest (i.e. natural) integrable systems that have a Lax form without spectral parameter. There are at least two possible reasons for this:

- either, it is known a priori that no curve should be present, for instance because the solutions are known to be exponentials (and not Abelian functions), which is for example the case for the non-periodic Toda lattices, as shown by Flaschka & Haine in [30] (see also Flaschka's lectures in [29]),
- or, the Lax form of the equation cannot be used to solve the equations and/or study the topology. This is the case e.g. for the Euler equations (see Chapter IV) and for

the Lax pair for the Kowalevski top constructed by Perelomov [69].

4. About this book

The reader will probably have understood that this book is devoted to the study of eigenvectors of Lax matrices and their use for topological purposes, mainly through examples coming from the mechanics of the rigid body.

I will not use much symplectic or Poisson geometry (the beginning of any textbook, e.g. Libermann & Marle [59] or Chapter II in the author's [10], will be sufficient), but a little bit more of algebraic geometry, curves, Jacobians, Riemann-Roch, Abel-Jacobi and a small amount of sheaf cohomology (among my favourite references are the books of Griffiths & Harris [37], Farkas & Kra [27] and Reyssat [78]) and of topology. I have tried to make the proofs as simple as possible.

4.1. What is in this book

I have chosen to show detailed examples of applications of the method discussed above: I have also tried to be exhaustive in the sense that there will be enough examples here to illustrate not only the questions in 2.2 but also the remarks in section 3. To take the reader directly to the heart of the subject, I have delayed all the "theory" to the appendices.

In the first chapter, things are set up: the differential equations of motion of the "top" are described as a Hamiltonian system. The case of a free (without gravity) rigid body and that of a heavy rigid body moving around its centre of mass are investigated in a classical and direct way. The case of the free rigid body is an example with only one degree of freedom, where the importance of the role played by algebraic curves (and by the fact that they are algebraic) is already very apparent.

Chapter II is devoted to the symmetric spinning top, a case studied by Lagrange in 1788 (see [56]). This is a rigid body with an axis of revolution. I will first recall that it is possible to express the solutions in terms of elliptic functions – and that this is what you see when you play with a spinning top.

After the classical treatments of the free rigid body and the Lagrange top comes the Lax pair/spectral curve/eigenvector method. Although the Lagrange top is a very old and well-known problem, even in our present viewpoint, it is actually not particularly simple – and the literature is often unclear[11] on the topological and/or real algebrogeometric aspects. I have tried to be rather precise and complete; it is even possible that some of the statements (e.g. II.2.4.1) are original.

The following chapter is of course, after that of the Euler-Poinsot and Lagrange cases, the study of the Kowalevski case. There are here several remarkable[12] features. First, it has been very difficult to find a Lax form with a spectral parameter for the differential system in this case. I have used here a Lax matrix in $\mathfrak{so}(3,2)$ manufactured by Bobenko, Reyman and Semenov-Tian-Shanski [18], but some others are available (those of Adler & van Moerbeke [4] and Haine & Horozov [40]). That of [18] has been obtained in a rather natural way and has a feature that is very important for us: the matrices are *real* so that the method works very well. It is known that the topology will be rather

[11] This is an euphemistic statement.

[12] First of all, the existence of three, and no more, cases. This question is briefly discussed in I.2.2.

complicated, but it is very well modelled, via the eigenvector mapping, by the Abelian surfaces[13] obtained; furthermore, this is the case where the method works best, although it is the most intricate. Last (?) surprise: it has been known since Kowalevski herself that the solutions can be expressed in terms of ϑ-functions associated with a genus-2 curve X, but the solutions produced by the Lax matrix of [18] come from ϑ-functions for the Prym variety of the covering $C \to E$ of a genus-1 curve by a genus-3 curve. As far as I know, the correspondences between these curves have not been completely clarified (this problem will also be considered in Appendix 5).

This chapter ends with some indications on the method used by Bobenko, Reyman & Semenov-Tian-Shanski [18] to find the Lax equation for the Kowalevski top, and with mention of two Lax equations related to the one used so far: the first describes once again the motion of a symmetric top and the second that of the Goryachev-Chaplygin top. The latter is only a curiosity but the first is there to show how the method depends on the Lax form used. This is also an example where the spectral curve is always singular.

Chapter IV is about the free rigid body. The results already demonstrated in Chapter I are re-explained via the eigenvector mapping (these results are due to Haine [39]) and a celebrated generalisation is investigated. The Lax equation here is due to Manakov [60] and seems to have been one of the bases for many of the developements explained here. Here the rigid body is four-dimensional, the system has two degrees of freedom, the first integrals are quadratic, the levels are both intersections of quadrics and affine parts of Abelian surfaces ... one can imagine that the algebro-geometric aspects are thus rich and interesting. A lot of papers have been devoted to these (e.g. those of Adler & van Moerbeke [3], Haine [39], Barth [17]). I will not consider here all these aspects. Neither will I give complete proofs for the topological questions. I will content myself with the determination of the regular levels and with the statement that the classical results of Haine (this is the "other approach" mentioned in 3.4) can give[14] all the desirable topological information.

The last chapter is devoted to an example in which the flows are not complete. We are forced to leave the world of rigid bodies there, as the total energy of a body is a proper function. The example here is an *ad hoc* variant of the periodic Toda lattice.

The appendices contain some of the technique needed. In the first appendix, we describe the natural Poisson structure on the dual of a Lie algebra and the Hamiltonian systems there. The second appendix is devoted to the Adler-Kostant-Symes (AKS) theorem, which allows us to construct many integrable systems in Lax form (this is the natural framework for all the examples presented here). In the third appendix, the definition of the eigenvector mapping is made precise and linearisation statements *à la* Griffiths are proved. In Appendix 4, some notions of algebraic geometry (complex and real) are set up: curves, Jacobians, real curves and their Jacobians. The last appendix is devoted to the definition and the study of some aspects of Prym varieties.

4.2. What is not in this book

- Although the examples come from mechanics, this is not a course of mechanics: I

[13] after a few manipulations!

[14] One of the most unsatisfying aspects of the state of the literature on the topology of integrable systems is well illustrated by, for instance, an announcement of the topological features of this example by Oshemkov [68], followed a few years later by the publication of some direct and tedious computations in the book [32] edited by Fomenko.

will explain only briefly how to get the differential system. The interested reader should have a look at the classical books of Jacobi [45], Klein & Sommerfeld [51], Appell [7], Whittaker [87] and Arnold [9], for instance.

- I have not written down the solutions of the equations, except in the cases of genus 1: there will be some $\wp$-functions in Chapters I and II. A complete treatment can be found in the beautiful text of Golubev [35]. The simplest way to write down the solutions in the Kowalevski case has been worked out by Bobenko, Reyman & Semenov-Tian-Shanski in [18].
- I have not tried to explain the word "integrability" (except for a few remarks in I.2.2), to discuss integrability "by quadratures", or action coordinates, although these are the subject of very interesting investigations (see e.g. the work of Vanhaecke [82]).
- Another very interesting question is how to find a Lax form *for a given system*. I know no satisfying answer to this question. Except for the constructions[15] of Bobenko, Reyman & Semenov-Tian-Shanski [18] (see also III.3), this seems to belong to the domain of cooking (see e.g. III.3.3).
- I have not tried to give an exhaustive bibliography. I hope however that the references given here together with the references they themselves contain will be sufficient. Mumford's book [66] should obviously be cited. Many aspects of the theory of integrable systems can be found in the recent book edited by Babelon, Cartier & Kosmann-Schwarzbach [16].
- All the algebro-geometric ideas illustrated here would seem to come from the work done by Novikov and others in the seventies on infinite-dimensional integrable systems (see Dubrovin's survey [26]). I have not tried to give a precise historical account of the material presented here. I send the interested readers to the historical notes in Reyman & Semenov-Tian-Shanski [77].

5. Notation

All the notation used here is as standard as possible. For instance $\mathcal{O}_C$ is the sheaf of holomorphic functions on C (usually a curve) and Ω^1_C that of holomorphic 1-forms. For a divisor D,

$$\mathcal{L}(D) = \{f \mid (f) + D \geq 0\}$$

and $h^0(D) = \dim \mathcal{L}(D)$.

If C is a smooth curve, $\operatorname{Jac}(C)$ is the quotient of the dual $H^0(\Omega^1_C)^*$ by the lattice Λ, the image of the integration map

$$H_1(C; \mathbf{Z}) \longrightarrow H^0\left(\Omega^1_C\right)^*,$$

while $\operatorname{Pic}(C)$ (resp. $\operatorname{Pic}^d(C)$) is the group of classes of linear equivalence of divisors (resp. degree-d divisors), or of isomorphism classes of line bundles (resp. degree-d line bundles); see Appendix 4.

Notice the difference between $\mathcal{O}_C$ and $\mathcal{O}_c$, the latter being an orbit labelled by a complex number c.

---

[15] The method used in [18] is the following: study all possible integrable systems that come from the Lie algebraic treatment explained in Appendix 2, and see what you get!

I

The rigid body with a fixed point

1. The equations

In order to begin to work, the first thing we need is a differential system. I will now explain rather quickly how to obtain it. There are numerous good books on this subject; the best known and most often cited are those of Appell [7], Whittaker [87], Golubev [35] and Arnold [9], but there is also a beautiful (Hamiltonian[1]) discussion in the Libermann & Marle book [59].

1.1. The differential system

We consider the motion of a rigid body about a fixed point O in a constant gravitational field γ. The configuration space is the group $SO(3)$ of rotations of the 3-space (this is to say that the body is rigid). We use a *moving frame* (relative frame, coordinates attached to the rigid body) and an absolute frame, both having their origin at O. To save notation, I shall use capital letters for vectors written in the moving frame, the corresponding lower-case letters denoting the same vector written in the absolute frame. If $R(t)$ is the rotation describing the moving frame as seen in the absolute frame at time t, we will have, for instance

$$q(t) = R(t)Q \tag{1}$$

to describe the motion of a point Q of the body and also

$$\Gamma(t) = R(t)^{-1}\gamma = {}^{t}R(t)\gamma \tag{2}$$

to describe the variation of the constant vector γ as viewed from the moving frame.

Take the derivative of relation (1) with respect to the time t:

$$\dot{q} = \dot{R}Q = \left(\dot{R}R^{-1}\right)q$$

[1]Surprisingly, the discussion of the spinning top in Arnold's book [9] is not "Hamiltonian".

As $R \in SO(3)$, $\dot{R}R^{-1}$ is a skew-symmetric mapping. Let us use now the isomorphism

$$(3) \qquad \begin{array}{ccc} \mathfrak{so}(3) & \xrightarrow{\varphi} & \mathbf{R}^3 \\ \begin{pmatrix} 0 & -z & y \\ z & 0 & -x \\ -y & x & 0 \end{pmatrix} & \longmapsto & \begin{pmatrix} x \\ y \\ z \end{pmatrix} \end{array}$$

which transforms the action of a matrix into a vector product:

$$A \cdot x = \varphi(A) \times x.$$

The avatar of the skew-symmetric $\dot{R}R^{-1}$ is a vector $\omega(t)$, the angular velocity, so that

$$\dot{q} = \omega \times q.$$

Let us now take the derivatives in equation (2) and use the skew-symmetry:

$$\dot{\Gamma} = {}^t\dot{R}\gamma = {}^t\dot{R}R\Gamma = -{}^tR\dot{R}\Gamma$$

and, still using the skew-symmetry,

$$\dot{\Gamma} = -{}^tR\left(\dot{R}{}^tR\gamma\right) = -{}^tR\left(\omega \times \gamma\right) = \left({}^tR\gamma\right) \times \left({}^tR\omega\right),$$

that is,

$$\dot{\Gamma} = \Gamma \times \Omega.$$

This expresses the three *Euler equations.* Recall that it is nothing more than a formulation – in the moving frame – of the fact that the gravitational field γ is constant.

Another series of three equations is obtained by consideration of the total angular momentum. Let q be a point of mass μ. Its (angular) momentum is $m = \mu q \times \dot{q} = \mu q \times (\omega \times q)$. Now the mapping $X \mapsto Q \times (X \times Q)$ is symmetric, i.e.

$$\begin{aligned} (Q \times (X \times Q)) \cdot Y &= \det(Q, X \times Q, Y) \\ &= \det(Y, Q, X \times Q) \\ &= (Y \times Q) \cdot (X \times Q), \end{aligned}$$

an expression which is symmetric in X and Y.

Averaging over all points of the rigid body, it is found that the total angular momentum M can be expressed as

$$M = \mathfrak{J}(\Omega)$$

where $\mathfrak{J}$, the inertia "matrix", is a positive-definite symmetric operator, as the previous computation shows, and is constant by definition. Roughly speaking, it describes the shape of the body. We shall investigate, for instance, the case where the body has an axis of revolution (containing the fixed point), the "symmetric" or Lagrange top: this geometrical property will appear as the fact that $\mathfrak{J}$ has a double eigenvalue for the plane orthogonal to the axis (see 2.1).

Remark. If the (nonnegative) eigenvalues of $\mathfrak{J}$ are denoted by λ_1, λ_2 and λ_3, it is an easy exercise (see e.g. Arnold's book [9]), using that the mass density is everywhere nonnegative, to show that these eigenvalues must satisfy the same equalities as the lengths of the edges of a triangle in the Euclidean plane ($\lambda_1 \leq \lambda_2 + \lambda_3$, and so on), with equalities only if the body is planar.

Let us come back to our momentum M: the derivative $\dot{m} = \dot{\widehat{RM}}$ must be the sum n of the momenta (with respect to O) of the forces applied to the body:

$$n = RN = \dot{\widehat{RM}} = \dot{R}M + R\dot{M} = R(\Omega \times M) + R\dot{M}$$

so that $N = \Omega \times M + \dot{M}$. Call G the centre of mass of the body and let L be the constant vector $\overrightarrow{OG}$. As the only force here is gravitational, $N = \Gamma \times L$ so that we finally obtain the following differential system, sometimes called the Euler-Poisson[2] equations,

$$\text{(E)} \qquad \begin{cases} \dot{\Gamma} &= [\Gamma, \Omega] \\ \dot{M} &= [M, \Omega] + [\Gamma, L] \end{cases}$$

in which I have replaced vector products by brackets of skew-symmetric matrices (the map φ in (3) being an isomorphism of Lie algebras).

1.2. (Co)adjoint orbits and first integrals

The total energy of the system is

$$H = \underbrace{\frac{1}{2} M \cdot \Omega}_{\text{kinetic}} \quad + \quad \underbrace{\Gamma \cdot L}_{\text{potential}}$$

and I am going to check that (E) is the Hamiltonian system associated with H. To give a precise meaning to this assertion, we need some more structure. The system (E) is a differential equation in the space $\mathbf{R}^3 \times \mathbf{R}^3$ of pairs (Γ, M) – recall that L is constant and $M = \mathfrak{J}(\Omega)$. Even without any understanding of the mechanics here, a quick look at the form of the equations gives us the idea of considering this vector space as

$$\mathbf{R}^3 \times \mathbf{R}^3 \cong \mathfrak{so}(3)[\varepsilon]/\varepsilon^2 = \mathfrak{g}$$

so that equation (E) becomes

$$\text{(E')} \qquad \dot{\overbrace{\Gamma + \varepsilon M}} = [\Gamma + \varepsilon M, \Omega + \varepsilon L].$$

Now (understanding the mechanics or not), $\mathfrak{g}$ is the Lie algebra of the Lie group that is the tangent bundle to $SO(3)$, or of the group of rigid motions in $\mathbf{R}^3$, that is, in either case, the semi-direct product $SO(3) \ltimes \mathbf{R}^3$ given by the standard $SO(3)$-action on $\mathbf{R}^3$. Now the adjoint action[3] of $SO(3)$ on $\mathfrak{so}(3)$ is by conjugation

$$\mathrm{Ad}_g \cdot X = gXg^{-1}$$

if we want to consider $X \in \mathfrak{so}(3)$ as a skew-symmetric matrix, or by the rotations

$$\mathrm{Ad}_g \cdot X = g(X)$$

[2] I have already mentioned that these first three equations are called the Euler equations. However the whole system (E) is in Lagrange's book [56].

[3] See Appendix 1.

if we prefer to imagine that $X \in \mathbf{R}^3$ is a vector (we are still using the isomorphism φ above). Now the adjoint action on our $\mathfrak{g}$ is

$$\text{(4)} \qquad \begin{aligned} \operatorname{Ad}_{(g,v)} \cdot (X + \varepsilon Y) &= g(X) + \varepsilon \left(g(Y) - (gXg^{-1})(v)\right) \text{ or} \\ &= gXg^{-1} + \varepsilon \left(gYg^{-1} - [gXg^{-1}, v]\right). \end{aligned}$$

Now, to identify (E′) with a Hamiltonian system, the only thing we need is an invariant non-degenerate symmetric bilinear form for which the gradient of H is

$$\nabla_{\Gamma+\varepsilon M} H = \Omega + \varepsilon L$$

(see Appendix 1). The symmetric form $\langle X + \varepsilon Y, X' + \varepsilon Y' \rangle = X \cdot Y' + X' \cdot Y$ has all the required properties: in effect

$$dH_{\Gamma+\varepsilon M}(X + \varepsilon Y) = \Omega \cdot Y + X \cdot L$$

since $M \mapsto M \cdot \Omega$ is a quadratic form.

We still need to identify the (co)adjoint orbits and/or the Casimir functions. Using the formulae in (4), it is easily checked that the orbit of $X + \varepsilon Y$ is depicted by $\|X\|^2$ and $X \cdot Y$. The functions $\|\Gamma\|^2$ and $\Gamma \cdot M$ are certainly "trivial" first integrals for (E). This was actually obvious from the beginning. The gravitational vector Γ can become variable in the moving frame, but its length cannot change! Thus $\|\Gamma\|^2$ is constant during the motion (and the constant will be unity for us). Similarly $\Gamma \cdot M$ is the momentum with respect to the vertical (i.e. the direction of the gravitational field) and must be constant.

The orbits

$$\mathcal{O}_c = \left\{ \Gamma + \varepsilon M \mid \|\Gamma\|^2 = 1 \text{ and } \Gamma \cdot M = c \right\}$$

are symplectic 4-manifolds, actually diffeomorphic with the tangent bundle TS^2 of the unit sphere in $\mathbf{R}^3$ by

$$\begin{array}{ccl} \mathcal{O}_c & \longrightarrow & TS^2 = \left\{ (\Gamma, M) \mid \|\Gamma\|^2 = 1 \text{ and } \Gamma \cdot M = 0 \right\} \\ \Gamma & \longmapsto & (\Gamma, M - c\Gamma). \end{array}$$

The system (E) is thus a Hamiltonian system. If we want to count first integrals: on any four-dimensional $\mathcal{O}_c$, we have already one constant of motion, namely the Hamiltonian H itself.

Notice also that $H : \mathcal{O}_c \to \mathbf{R}$ is a proper mapping so that all the flows will be complete.

2. The question of integrability

In order to have an integrable system in the sense explained in the Introduction, we need another first integral commuting with the energy H.

There is a very classical list[4] of cases (depicted by specific values of L and $\mathcal{J}$) in which such an integral is known. I will write down this list (this is unavoidable) and make a few comments on the significance it can have to do so.

[4] A discussion of the relative contributions of the people mentioned in this list and complete references with dates (of course, Poinsot did nothing in 1758!) can be found in the classical book of Whittaker [87].

2.1. A list

The Euler-Poinsot case (1758). Here the fixed point is the centre of mass $O = G$ or $L = 0$. Obviously $K = \frac{1}{2}\|M\|^2$ is a constant of motion. In effect

$$\nabla_{\Gamma+\varepsilon M} H = \Omega \quad \text{and} \quad \nabla_{\Gamma+\varepsilon M} K = M$$

so that[5]

$$\{H, K\}(\Gamma + \varepsilon M) = \langle \Gamma + \varepsilon M, [\Omega, M] \rangle = M \cdot (\Omega \times M) = 0.$$

The Lagrange case (1788). This is the case of the symmetric top, where there exists a (moving) frame in which the inertia matrix is

$$\mathcal{J} = \begin{pmatrix} l & 0 & 0 \\ 0 & l & 0 \\ 0 & 0 & m \end{pmatrix}$$

(up to a change of units, one can – and I will – assume that $l = 1$) and whose third vector is collinear to L: the line OG is an axis of revolution of the body. Notice that this means that $(M - \Omega) \times L = 0$.

This case is (universally) called the Lagrange case. The momentum $K = M \cdot L$ with respect to the axis (the Lagrange momentum) is a first integral. Moreover $\nabla_{\Gamma+\varepsilon M} K = L$ and

$$\begin{aligned} \{H, K\}(\Gamma + \varepsilon M) &= \omega_{\Gamma+\varepsilon M}(\Omega + \varepsilon L, L) \\ &= (\Gamma + \varepsilon M) \cdot [\Omega + \varepsilon L, L] \\ &= (\Gamma + \varepsilon M) \cdot [\Omega, L] \\ &= M \cdot (\Omega \times L) \\ &= 0 \end{aligned}$$

since M is in the vector subspace spanned by Ω and L, so that the two integrals do commute.

Notice also that K is the momentum for a flow of rotations about the symmetry axis L. Its Hamiltonian vector field is actually

$$\begin{aligned} X_K(\Gamma + \varepsilon M) &= [\nabla_{\Gamma+\varepsilon M} K, \Gamma + \varepsilon M] \\ &= [L, \Gamma + \varepsilon M] \\ &= [L, \Gamma] + \varepsilon [L, M]. \end{aligned}$$

The vector $[L, X] = L \times X$ is obtained by the projection of X onto the plane orthogonal to L followed by a rotation of a quarter of turn: this is indeed the fundamental vector field of the S^1-action by rotations about the axis L. The function K is a periodic Hamiltonian.

The Kowalevski case (1889). There exists a moving frame in which the inertia matrix is

$$\mathcal{J} = \begin{pmatrix} 2m & 0 & 0 \\ 0 & 2m & 0 \\ 0 & 0 & m \end{pmatrix}$$

[5]See Appendix 1 for the definition of the Poisson bracket $\{\ ,\ \}$ and the way to compute it in terms of gradients.

(we will assume that $m = 1$) and whose *first* vector is collinear with L: there is an "equatorial" plane, as in the symmetric case, but it contains the centre of mass. This is the Kowalevski case [55]. Let us write, in the same basis,

$$M = \begin{pmatrix} u \\ v \\ w \end{pmatrix}, \quad \Omega = \begin{pmatrix} p \\ q \\ r \end{pmatrix}, \quad \Gamma = \begin{pmatrix} \gamma_1 \\ \gamma_2 \\ \gamma_3 \end{pmatrix}, \quad L = \begin{pmatrix} -1 \\ 0 \\ 0 \end{pmatrix}$$

and consider the function

$$K = \left|(p+iq)^2 + (\gamma_1 + i\gamma_2)\right|^2.$$

It is quite easy to check (by a direct computation) that $\{H, K\} = 0$ so that the system (E) will be integrable in this case. Since this would not be much fun and in any case will be a consequence of what follows (namely the AKS theorem, see Appendix 2 and the constructions in Chapter III), it is left as an exercise for the reader.

The Goryachev-Chaplygin case (1900). This is a slightly different case[6] because there is only one orbit, namely the orbit $\mathcal{O}_0$, on which the system is integrable. In other words, it is not integrable as a system on the whole Poisson manifold, but only on a single symplectic leaf. This is the case where

$$\mathcal{J} = \begin{pmatrix} 4m & 0 & 0 \\ 0 & 4m & 0 \\ 0 & 0 & m \end{pmatrix}$$

again in a basis whose first vector is collinear to L. Using the same notation as in the Kowalevski case for the coordinates of M, Ω, Γ and L, the second first integral is

$$K = w\left(u^2 + v^2\right) + 2u\gamma_3.$$

Here again, a direct computation gives

$$\nabla_{\Gamma+\varepsilon M} K = \begin{pmatrix} 2uw + 2\gamma_3 \\ 2vw \\ u^2 + v^2 \end{pmatrix} + 2\varepsilon \begin{pmatrix} 0 \\ 0 \\ u \end{pmatrix},$$

and then

$$\{H, K\}(\Gamma + \varepsilon M) = cv$$

so that the system is integrable on the orbit $\mathcal{O}_0$.

Remark. We have already noticed that the total energy H is a proper mapping $\mathcal{O}_c \to \mathbf{R}$ so that all the common levels of the first integrals are compact in all these cases: indeed, they consist of Liouville *tori*.

[6]The references to the original papers of Goryachev and Chaplygin can be found in Golubev's book [35].

2.2. Comments

One could (should?) wonder how it is possible to find these cases and nothing more. This is a very interesting problem. The method used by Kowalevski is the following. She investigated when (looking for a condition on $\mathcal{J}$ and L) the singularities of the solutions of the system (E) are simple enough. More precisely, the system (E) being nonlinear, its solutions can have very complicated singularities; the simplest thing one can expect is that the solutions are *meromorphic* functions of time. At the time of Kowalevski, the solutions were well known in the Euler-Poinsot and Lagrange cases. She noticed that they were meromorphic and looked for all the possible inertia matrices that would provide such a behaviour. Besides the old cases, she found what is now called the Kowalevski case, and noticed that the system had an additional integral (K as mentioned above) and moreover gave the solutions.

Remarks

1. As we have already mentioned at least twice, the total energy is proper, so that nothing can go to infinity. In order to think of the solutions as meromorphic functions (or worse), it was at least necessary to consider time as a complex variable. This is certainly why this was not done before the end of the nineteenth century, and why it was done by a mathematician of Weierstrass' school.
2. There is much modern investigation into the problem raised by this procedure, namely why should a differential equation with not-too-singular solutions have a lot of conserved quantities? This still seems to be rather mysterious.
3. A related and very active research field is the so-called "Painlevé analysis" (or, more accurately, "Kowalevski-Painlevé analysis") where the *poles* of the (meromorphic) solutions are used to understand the geometry of the level sets (a few references will be given in Chapters IV and V).

About the completeness of the list above: it can be shown that, if the system (E) has an additional polynomial (resp. meromorphic) integral, then we are in one of the cases above. This was achieved by Husson [44] (resp. Ziglin [89]).

3. The three-dimensional free rigid body and the Euler-Poinsot case

3.1. The free rigid body

In the case where the fixed point is the centre of mass (the Euler-Poinsot case), the differential system (E) becomes

$$\begin{cases} \dot{M} &= [M, \Omega] \\ \dot{\Gamma} &= [\Gamma, \Omega] \\ M &= \mathcal{J}(\Omega) \end{cases}$$

since $L = 0$. Of course, the main problem is to solve the first equation (the solution of the second will follow). We thus have to consider the system

$$\begin{cases} \dot{M} &= [M, \Omega] \\ M &= \mathcal{J}(\Omega), \end{cases}$$

which describes the motion of a *free* rigid body (no forces, not even gravity) about one of its points. This is certainly a good example to start with, because it has very low dimension: $M \in \mathfrak{so}(3)$ where the orbits are the spheres centred at 0; any Hamiltonian system on a surface is integrable.

The "second first integral" $K = \frac{1}{2}\|M\|^2$ that we have exhibited for the whole system becomes here a "trivial" integral, one that describes the orbits (spheres)

$$\mathcal{O}_{p^2} = \left\{M \in \mathbf{R}^3 \mid u^2 + v^2 + w^2 = 2p^2\right\},$$

the symplectic manifolds on which we will investigate the levels of the Hamiltonian (here the kinetic energy)

$$H = \frac{1}{2}\left(\lambda_1 u^2 + \lambda_2 v^2 + \lambda_3 w^2\right).$$

Here λ_1, λ_2 and λ_3 are nonnegative real numbers (inverses of the eigenvalues of $\mathcal{J}$). I will assume that they are distinct and that, say, $\lambda_1 < \lambda_2 < \lambda_3$ so that the levels of H are general ellipsoids (i.e. without an axis of revolution) and we are looking at their intersection with the spheres $\mathcal{O}_{p^2}$. This is easily described and is depicted in Figure 1. The regular levels consist of two disjoint ovals. The critical levels consist either of two points or of two ovals intersectiong at two points. The shaded zones correspond to empty levels, the half-lines to critical values.

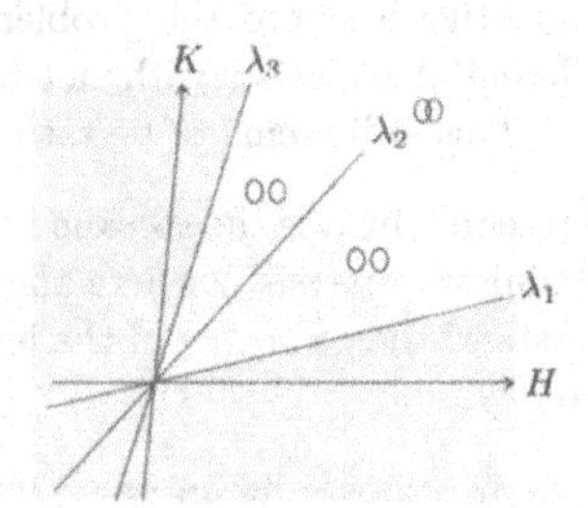

Figure 1: Regular levels for the free rigid body.

3.2. An algebraic model

All the connected components of a regular level are topologically circles. In this dimension, the Arnold-Liouville theorem is not really needed to obtain this conclusion. The fact that the Hamiltonian vector field X_H is linear once the levels are identified with circles is more or less tautological until we are able to understand the affine structure of the levels.

Already, here, we have an algebraic structure that allows us to give an elegant description of the whole situation. Our level is the intersection of a sphere and an ellipsoid, that is, of two real affine quadrics. Nothing prevents us from considering them to be complex and projective, so that their equations are written

$$\begin{cases} u^2 + v^2 + w^2 - 2p^2t^2 = 0 \\ \lambda_1 u^2 + \lambda_2 v^2 + \lambda_3 w^2 - 2ht^2 = 0 \end{cases}$$

Call V this intersection. It is a classical result in algebraic geometry that such an intersection of quadrics is an elliptic curve.

3.2.1 PROPOSITION. *If λ_1, λ_2 and λ_3 are mutually distinct and also distinct from h/p^2, the intersection V is a smooth genus-1 curve in $\mathbf{P}^3(\mathbf{C})$, the double covering of $\mathbf{P}^1(\mathbf{C})$ branched at λ_1, λ_2, λ_3 and h/p^2.*

Remark. This is the most general situation of two smooth transverse quadric surfaces (at least over $\mathbf{C}$).

Proof. The condition imposed on the coefficients ensures that the two quadric surfaces are transversal. The intersection V is thus a smooth complex curve. Consider now the linear family (i.e. the pencil) $\mathcal{L}$ consisting of all the quadrics C_z:

$$C_z: \quad Q_z(u,v,w,t) = (\lambda_1 - z)\,u^2 + (\lambda_2 - z)\,v^2 + (\lambda_3 - z)\,w^2 - 2\left(h - zp^2\right)t^2 = 0,$$

where $z \in \mathbf{P}^1 = \mathbf{C} \cup \infty$ so that C_0 corresponds to our ellipsoid and C_∞ to the sphere of radius $p\sqrt{2}$. Notice that the quadrics of the pencil $\mathcal{L}$ that are singular correspond to $z = \lambda_1, \lambda_2, \lambda_3$ or h/p^2.

Let us now fix a point $A \in V$. This gives a mapping $\Phi_A : V \to \mathcal{L}$, which I will now describe. Every point $M \in V$ defines a line AM (the tangent to V at A if $M = A$); this line is contained in one (and only one) quadric C_z: if φ_z is the symmetric bilinear form associated with Q_z (its polar form),

$$Q_z\left[(1-\alpha)A + \alpha M\right] = 0 \quad \forall \alpha \Leftrightarrow \varphi_z(A, M) = 0.$$

This is an equation of degree 1 in the variable z and thus gives a unique z such that $AM \subset C_z$.

The mapping Φ_A is holomorphic and I am going to show that it is the double covering of $\mathbf{P}^1$ announced in the statement of the proposition. Let us choose a point $z \in \mathbf{P}^1$ and look at its inverse image $\Phi_A^{-1}(z)$. It consists of all straight lines in C_z that contain A, that is, of the intersection $T_A C_z \cap Q_z$; there are two of these lines. Let us intersect the pencil with the plane $T_A C_z$. Of course, what we get is a pencil of conic curves. Our two lines form a singular conic of this pencil, so that they must intersect all the conics of the pencil; hence we get two points M, $N \in V$ of which z is the image (see Figure 2).

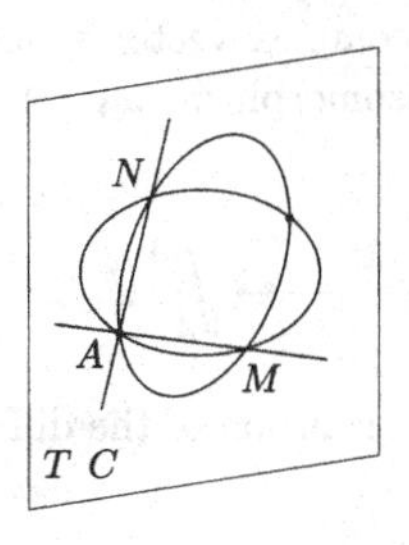

Figure 2

Now these two points will be distinct unless the two lines coincide, that is, unless $T_A C_z \cap Q_z$ is a double line, which means that C_z is singular, equivalently that $z = \lambda_1$, λ_2, λ_3 or h/p^2.

Thus we do have a double cover, branched at four points, and V is indeed a genus-1 curve. □

A remark on the real part. Assume the point A we have used to define the covering map Φ_A is a real point (has real coordinates). Then Φ_A is a real map and sends the real part $V_{\mathbf{R}}$ of V to $\mathbf{P}^1(\mathbf{R})$. The image is the set of all the real values z for which C_z contains a *real* line through A. Now the real quadric surface C_z can contain a real line only if it is a one-sheeted hyperboloid, that is, if the signature of the quadratic form Q_z is $(2,2)$, or if that of

$$\frac{\lambda_1 - z}{h - zp^2}u^2 + \frac{\lambda_2 - z}{h - zp^2}v^2 + \frac{\lambda_3 - z}{h - zp^2}w^2$$

is $(2,1)$. If we put the four points $\{\lambda_1, \lambda_2, \lambda_3, h/p^2\}$ in the increasing order $\lambda_1 < b < c < \lambda_3$, the condition is that

$$z \in [\lambda_1, b] \cup [c, \lambda_3].$$

Notice that I have used a real point A of V so that I must have assumed that the real part $V_{\mathbf{R}}$ was non-empty and thus that $h/p^2 \in [\lambda_1, \lambda_3]$ (this is why the increasing order above started with λ_1 and ended with λ_3). Then $V_{\mathbf{R}}$ has two connected components as we have already noticed.

3.3. Linearisation of the solutions

Let us now write the differential equation $\dot{M} = [M, \Omega]$ in components, that is:

$$\begin{cases} \dot{u} &= (\lambda_3 - \lambda_2)\, vw \\ \dot{v} &= (\lambda_1 - \lambda_3)\, wu \\ \dot{w} &= (\lambda_2 - \lambda_1)\, uv \end{cases}$$

so that we must consider the differential form

$$\omega = \frac{du}{(\lambda_3 - \lambda_2)\, vw} = \frac{dv}{(\lambda_1 - \lambda_3)\, wu} = \frac{dw}{(\lambda_2 - \lambda_3)\, uv}.$$

It is an easy exercise (left to the reader) to check that this is a holomorphic 1-form on V. As V is a genus-1 curve, the complex vector space of holomorphic 1-forms on V is one-dimensional and there is an isomorphism $u_A : V \to \mathbf{C}/\Lambda$ (where Λ is the period lattice of V) by integration:

$$M \longmapsto \int_A^M \omega.$$

Suppose now that $t \mapsto M(t)$ is a solution of the differential equation $\dot{M} = [M, \Omega]$. We have

$$u_{M(0)}(M(t)) = \int_{M(0)}^{M(t)} \omega = \int_{M(0)}^{M(t)} dt = t.$$

This means that the images of the solutions in $\mathbf{C}/\Lambda$ are straight lines $\bmod \Lambda$ (in a dimension-1 space this is not really overwhelming) with a *linear* parametrisation (this is the interesting point). Thus the canonical affine structure of $\mathbf{C}/\Lambda$ is a reification of the affine structure given by the flow on V.

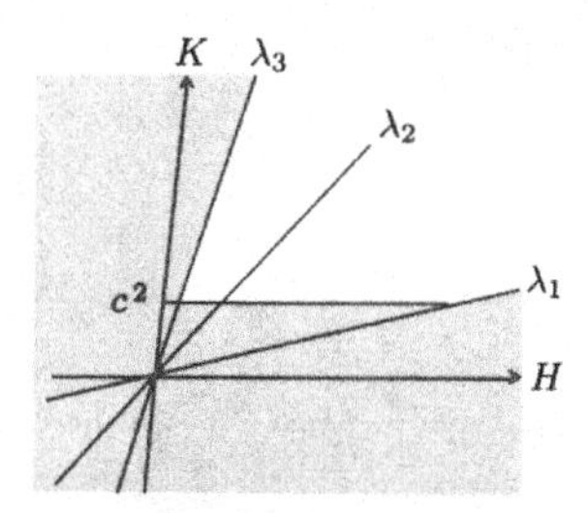

Figure 3: Momentum mapping for the Euler-Poinsot top.

3.4. Liouville tori for the Euler-Poinsot top

Let us come back to our initial Euler-Poinsot problem and take care of Γ. For any M satisfying the equations

$$\begin{cases} K = \frac{1}{2}p^2 = \frac{1}{2}\|M\|^2 \\ H = \frac{1}{2}h = \frac{1}{2}M \cdot \Omega \end{cases}$$

we are looking for all the vectors Γ such that

$$\begin{cases} \|\Gamma\|^2 = 1 \\ \Gamma \cdot M = c. \end{cases}$$

The second equation is that of a plane orthogonal to M and we are looking for the intersection of this plane with the unit sphere. This

- is empty if $|c| > p/\sqrt{2}$
- is a point ($\Gamma = M$) if $|c| = p/\sqrt{2}$
- is a circle if $|c| < p/\sqrt{2}$.

Combining these remarks with what we have found for M, we derive the image of the "momentum mapping" (H, K) (Figure 3) and the fact that all the regular levels consist of two Liouville tori.

The critical levels corresponding to (generic) points of the boundary of the image consist of two circles; those corresponding to the slope-λ_2 line are the product of two intersecting circles with a circle.

II

The symmetric spinning top

1. Introduction to the symmetric spinning top

I will first explain in this section, in a very classical way (see e.g. Appell [7] and Arnold [9]) the motion of the axis of revolution of a spinning top (anyone who has ever played with a true spinning top understands that this is interesting). I will also explain the relation of this motion to the problem of Liouville tori.

In this case, the inertia matrix has the form

$$\mathcal{J} = \begin{pmatrix} 1 & 0 & 0 \\ 0 & 1 & 0 \\ 0 & 0 & m \end{pmatrix}$$

as above (see I.2.1 for the notation that we use here) and the axis L is the third vector of the basis. Let us fix an orbit $\mathcal{O}_c$ and values h and k of the quantities $H = \frac{1}{2} M \cdot \Omega + \Gamma \cdot L$ and $K = M \cdot L$.

1.1. The differential equation for γ_3

To look at the position of the axis L with respect to the vertical direction Γ amounts to looking at the quantity $M \cdot \Gamma$. Write

$$\Gamma = \begin{pmatrix} \gamma_1 \\ \gamma_2 \\ \gamma_3 \end{pmatrix}, \quad M = \begin{pmatrix} u \\ v \\ w \end{pmatrix}, \quad \Omega = \begin{pmatrix} u \\ v \\ m^{-1} w \end{pmatrix}$$

and look at γ_3. Let us try to find the differential equation it satisfies.

Since $\dot{\Gamma} = [\Gamma, \Omega]$, we have $\dot{\gamma_3} = \gamma_1 v - \gamma_2 u$ and

$$\dot{\gamma_3}^2 = \gamma_1^2 v^2 + \gamma_2^2 u^2 - 2\gamma_1 \gamma_2 uv. \tag{1}$$

Let us use now that $w = M \cdot L = k$ and $c = M \cdot \Gamma$. This gives

$$u\gamma_1 + v\gamma_2 = c - k\gamma_3,$$

which we can square to compute the term $2uv\gamma_1\gamma_2$, and then replace in equation (1):

$$\dot{\gamma_3}^2 = \left(\gamma_1^2 + \gamma_2^2\right)\left(u^2 + v^2\right) - \left(c - k\gamma_3\right)^2.$$

Since Γ is a unit vector, we get

$$\dot{\gamma_3}^2 = \left(1 - \gamma_3^2\right)\left(u^2 + v^2\right) - \left(c - k\gamma_3\right)^2. \tag{2}$$

To get rid of the terms in u and v that still remain in equation (2), we can now use the last integral, H:

$$h = \frac{1}{2}\left(u^2 + v^2 + \frac{1}{m}k^2\right) + \gamma_3$$

thanks to which it is easy to compute $u^2 + v^2$, which we replace in (2), writing

$$H' = 2H + \left(1 - \frac{1}{m}\right)K^2.$$

We get, at last:

$$\dot{\gamma_3}^2 = \left(1 - \gamma_3^2\right)\left(h' - k^2 - 2\gamma_3\right) - \left(c - k\gamma_3\right)^2. \tag{3}$$

Now, equation (3) has the form $\dot{x}^2 = f(x)$ for some degree-3 polynomial f. We thus have to consider the curve C of the equation

$$y^2 = f(x) \tag{4}$$

... an elliptic curve again, on which equation (3) is equivalent to $dt = dx/y$.

In order that the values (h, k) correspond to real motions of the spinning top, there must be solutions with $-1 \le \gamma_3 \le 1$, so that the curve must have real points over the interval $[-1, 1]$. The direct computation leading to the values of (h, k) allowed by this condition is rather tedious and we will obtain the same result quite easily in 2.5, so that now I will just assume that the condition is fulfilled.

Hence, as $f(\pm 1) = -(c - K)^2 \le 0$, the polynomial f must have two real roots x_3 and x_2 in $[-1, 1]$ and a third real root $x_1 > 1$.

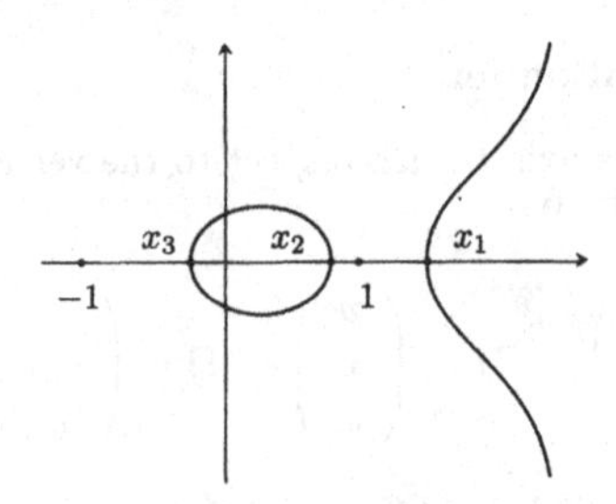

Figure 4: The curve C.

The real part of C is depicted in Figure 4. Notice that the bounded component is the only one that corresponds to real motions. This also means that the solutions of the differential equation (3) that correspond to real motions of the top are confined to $[x_3, x_2]$, so that the end of the axis will stay in the zone delimited by x_2 and x_3 on the sphere as one is expecting and as can be seen in the classical pictures in Figure 7.

1.2. Symplectic reduction and the Liouville tori

Since the Lagrange case was defined by the existence of an axis of revolution, there is a circle group acting on the phase space. Look at a symplectic[1] orbit $\mathcal{O}_c$ and consider the Lagrange momentum

$$K : \mathcal{O}_c \longrightarrow \mathbf{R}.$$

Fix a regular level $\mathcal{O}_{c,k} = K^{-1}(k)$ of K. Since the group $\mathcal{R}$ of rotations about the axis L is generated by the flow of K (we have already noticed this in I.2.1), it still acts on $\mathcal{O}_{c,k}$ so that we can consider the quotient $\mathcal{O}_{c,k}/\mathcal{R}$. This is a simple example of *symplectic reduction*, a process invented by Marsden & Weinstein [61] (see also any textbook on symplectic geometry, e.g. Libermann & Marle [59] or the author [10]): the kernel of the restriction of the symplectic form to the codimension-1 submanifold $\mathcal{O}_{c,k}$ is generated by the Hamiltonian vector field X_K, so that it defines a non-degenerate closed 2-form on the quotient, which is thus a dimension-2 symplectic manifold. Moreover, as the energy commutes with K, it defines a Hamiltonian, still called H,

$$H : \mathcal{O}_{c,k} \longrightarrow \mathbf{R}.$$

Now a common level $\mathcal{T}_{h,k} \subset \mathcal{O}_c$ of the two integrals is an S^1-bundle over a level of H in $\mathcal{O}_{c,k}/\mathcal{R}$, so that the reduction allows us to understand the "Liouville tori" quite easily. As in the case of the free rigid body, we only need to look at the levels of a function H on a surface.

First, let us look at the $\mathcal{R}$-orbit of a point (Γ, M) of $\mathcal{O}_{c,k}$. The circle $\mathcal{R}$ acts by the same rotation on the plane vectors (γ_1, γ_2) and (u, v). Notice that the points (Γ, M) with $\gamma_1 = \gamma_2 = 0$ are critical points of K in $\mathcal{O}_c$ (easily verified) so that, since we have assumed that k is a regular value, any orbit has a unique representative such that $\gamma_1 > 0$ and $\gamma_2 = 0$. Once we have chosen a γ_3, then $\gamma_1 = \sqrt{1 - \gamma_3^2}$, the equation $\gamma_1 u + \gamma_3 k = c$ gives $u = (c - k\gamma_3)/\gamma_1$ and v can take any value. Any $\mathcal{R}$-orbit is detected by a γ_3 and a v, so that we get an embedding of $\mathcal{O}_{c,k}/\mathcal{R}$ into $]-1, 1[\times \mathbf{R}$.

Now, we can look at the levels of H:

$$H(\gamma_3, v) = \frac{1}{2}\left[\left(\frac{c - k\gamma_3}{\gamma_1}\right)^2 + v^2 + \frac{1}{m}k^2\right] + \gamma_3$$

so that, at level h, we get

$$v^2 = (1 - \gamma_3)^2 \left(h' - k^2 - 2\gamma - 3\right) - (c - k\gamma_3)^2$$

... our point (γ_3, v) must belong to the elliptic curve $\mathcal{C}$. Thus we have proved:

1.2.1 PROPOSITION. *All the regular levels in $\mathcal{O}_c$ of the momentum mapping for the symmetric top are circle bundles over a connected component of a real elliptic curve. In particular, they are tori (and thus connected).* □

---

[1] One can describe the reduction (quotient) process in the whole Poisson manifold. As it is easier to restrict a 2-form to a submanifold than to restrict a Poisson bracket, we will feel more comfortable in the symplectic framework and take the quotient of each orbit individually.

1.3. A remark on the Liouville tori

Here is another (direct) way of seeing that the regular levels are connected in this example. Fix a value (h, k) such that the corresponding level is non-empty: we get two real numbers x_3 and x_2 such that $-1 \leq x_3 \leq x_2 \leq 1$. It is certainly true that we will even have, for a generic (h, k), $-1 < x_3 < x_2 < 1$. Fix a unit vector Γ such that $x_3 \leq \gamma_3 \leq x_2$ and look at all the vectors M such that $(\Gamma, M) \in \mathcal{O}_c$ lies in the (h, k) level in $\mathcal{O}_c$. These are the vectors

$$\begin{pmatrix} u \\ v \\ k \end{pmatrix} \text{ such that } \begin{cases} u\gamma_1 + v\gamma_2 &= c - k\gamma_3 \\ \frac{1}{2}\left(u^2 + v^2\right) &= -\frac{1}{2m}k^2 - \gamma_3 + H. \end{cases}$$

This set of vectors is the intersection of a line and a circle in the (u, v) plane. It is non-empty by assumption and consists, in general, of two points. Thus the level is *one* torus, which the projection $(\Gamma, M) \mapsto \Gamma$ sends to an annulus (Figure 5), which is our zone on the sphere.

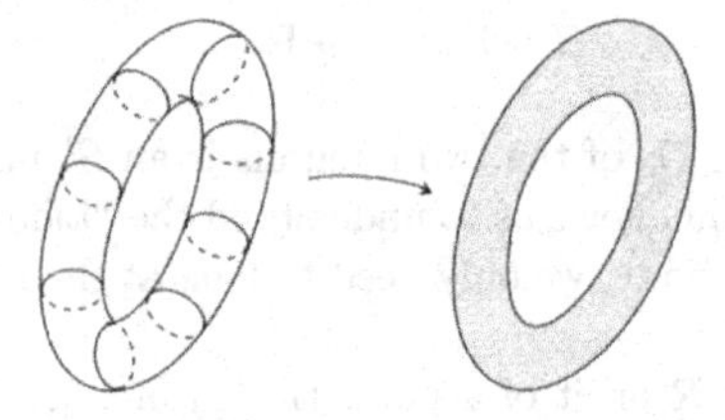

Figure 5

Rather often, it is possible to visualise that a system is integrable because one has appreciated that the solutions stay in annuli. This is the case for the spinning top with annulus $x_3 \leq \gamma_3 \leq x_2$ on the 2-sphere, as is apparent in Figure 7, but also for geodesics on an ellipsoid (see Hilbert & Cohn-Vossen [41], Knörrer [52] or the author [11], and Figure 6).

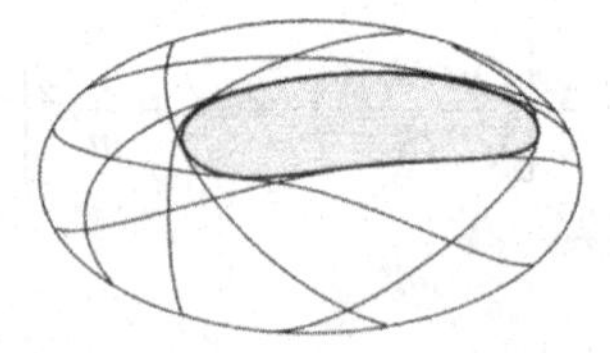

Figure 6: Geodesics on the ellipsoid.

1.4. The solutions of the differential eqution for γ_3

The differential equation (3) can be written $dt = dx/y$ (here x is γ_3 and $y^2 = f(x)$). A small (real) change of variables $X = ax + b$, $Y = y$ brings the equation of $\mathcal{C}$ to the standard form

$$Y^2 = 4X^3 - g_2X + g_3$$

so that the complex curve is parametrised by a Weierstrass $\wp$-function and its derivative:

$$X = \wp(u), \quad Y = \wp'(u)$$

and the solutions of the differential equation $du = dX/Y$ are the functions $X(u) = \wp(u + \text{constant})$.

As for the rewritten equation (3),

$$dt = \frac{dx}{y} = \frac{1}{a}\frac{dX}{Y} = \frac{1}{a}du,$$

its solutions are $X = \wp(at + \text{const})$ and thus

$$\gamma_3(t) = x(t) = \frac{1}{a}\left[\wp\left(at + \text{const}\right) - b\right]. \tag{5}$$

It is quite easy to compute a and b explicitly: the translation takes the sum of the roots of f to 0 and the dilation produces the coefficient 4. Thus (5) actually gives all the *complex* solutions of (3).

We have already noticed that the real part of $\mathcal{C}$ has two connected components. As a complex torus, the curve $\mathcal{C}$ (or rather its completion, we must add a point at infinity) has the form $\mathbf{C}/\Lambda$ for a certain lattice Λ. Because of the real properties of $\mathcal{C}$, it can be assumed that Λ has a basis $(2\omega, 2\omega')$ with $2\omega \in \mathbf{R}$ and $2\omega' \in i\mathbf{R}$, and that the real structure is given by the complex conjugation $z \mapsto \bar{z}$. The two components of the real part appear, in this framework, as the images of the real axis and of the parallel line through ω' (see Appendix 4 and especially Figure 25).

The Weierstrass $\wp$-function has double poles at the lattice points, so that, for $u \in \mathbf{R}$, $\wp(u)$ parametrises the non-compact real component. As for the compact component, it is parametrised by the $\wp(u + \omega')$ ($u \in \mathbf{R}$) – which are also real. The $\wp$-function has no pole on this line, and is an honest periodic function (see Figure 26).

1.5. The motion of the axis

For the sake of completeness – and also because it is beautiful – let us recall how the motion of the axis can be described. We have defined γ_3 as the component of Γ along the axis of the top, but we can as well think of it as the component of the vector L on the vertical axis: we are now in the absolute frame. From this point of view, what we have just described is the so-called *nutation* motion (oscillations with respect to the vertical). As we have explained, the end of the axis is confined in a zone bounded by two parallels ($x_3 \leq \gamma_3 \leq x_2$) on the unit sphere and oscillates periodically.

The complete motion of the top consists of this nutation, of a rotation around its axis and of a *precession* motion that I intend to describe now. Decompose the rotation R relating the absolute and relative frames (that of I.1) as the product of three rotations, using the Euler[2] angles θ, φ and ψ.

- θ is the angle between the axis and the vertical direction, so that $\gamma_3 = \cos\theta$.
- The equatorial plane of the top (the plane $L^\perp$) intersects the horizontal plane along a line OI and ψ is the angle between the axis OX of the moving frame and this line. This is what the astronomers call the *azimuth*. It is the variation of this angle that we want to study now.

[2]The notation I am using here is that of Appell's book [7].

- To take the axis OI to the axis Ox of the absolute frame, rotate it around the vertical. The last angle, φ, is the angle of this rotation.

Now, it is not very hard to check that

$$\Omega = \begin{pmatrix} \dot\psi \sin\theta \sin\varphi + \dot\theta \cos\varphi \\ \dot\psi \sin\theta \cos\varphi - \dot\theta \sin\varphi \\ \dot\psi \cos\theta + \dot\varphi \end{pmatrix}$$

(see e.g. Arnold's book [9]), so that

$$\dot\psi = \frac{c - k\gamma_3}{1-\gamma_3^2}.$$

Hence, according to the position of c/k with respect to the two roots x_3 and x_2 of f, the function $\psi(t)$ will or will not be monotonic. The three diagrams in Figure 7 show, from left to right, the cases $c/k \notin [x_3, x_2]$, $c/k = x_2$ (the case $c/k = x_3$ is left as an exercise for the reader) and $c/k \in]x_3, x_2[$.

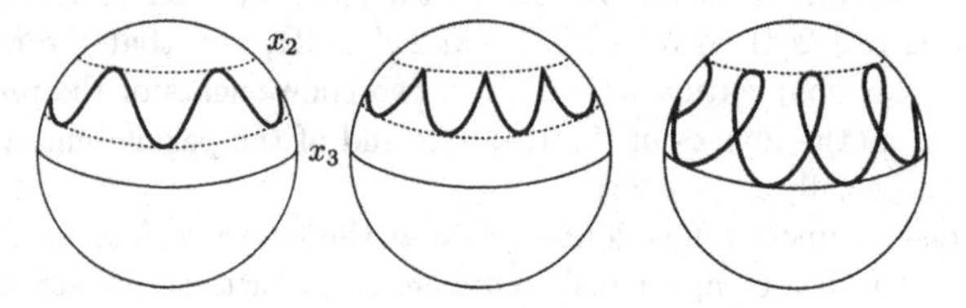

Figure 7: The motion of the axis.

2. A Lax pair and what follows

Here begins the use of the technique described in the Introduction, and this is also a good time to send the reader back to Verdier's exposition [84] where this technique is explained starting from the "simple"[3] example of the symmetric top. The very same example is investigated or mentioned in every paper on integrable systems, in particular, in the classical Adler & van Moerbeke [2] (I believe that the Lax equation used below originated in this paper) and in Ratiu & van Moerbeke [73].

2.1. The Lax equation

Still using the notation of Chapter I and more precisely of I.2.1, we have:

2.1.1 PROPOSITION. *The differential system for the motion of the Lagrange top is equivalent to*

$$\overbrace{\Gamma\lambda^{-1} + M + L\lambda} = \left[\Gamma\lambda^{-1} + M + L\lambda, \Omega + \lambda L\right].$$

[3] I will explain the quotation marks in 2.4.

Proof. This equation was obtained from equation (E′) of Chapter I, namely

$$\widehat{\Gamma+\varepsilon M}=[\Gamma+\varepsilon M,\Omega+\varepsilon L]$$

by replacing ε by an indeterminate λ (remember that $\varepsilon^2=0$) and shifting the exponents to make the equation look nicer. The only thing we need to do is to check that the coefficients of λ^2 and λ on the two sides agree. The coefficient of λ^2 is obviously zero on both sides. As L is constant, the linear term is zero on the left-hand side. Looking at the right-hand side, we need only to prove that $[L,\Omega]+[M,L]=0$, that is, $(M-\Omega)\times L=0$. Our assumption on the form of the inertia matrix $\mathfrak{J}$ is precisely that $M-\Omega$ is collinear with L. Hence the proposition. □

Having in mind the so-called AKS theorem (see Appendix 2), we could consider the following differential equations:

$$\frac{d}{dt}\left(\Gamma\lambda^{-1}+M+L\lambda\right)=\left[\Gamma\lambda^{-1}+M+L\lambda,M+L\lambda\right],$$

where the second matrix $M+L\lambda$ is the polynomial part of $\Gamma\lambda^{-1}+M+L\lambda$, and

$$\frac{d}{dt}\left(\Gamma\lambda^{-1}+M+L\lambda\right)=\left[\Gamma\lambda^{-1}+M+L\lambda,L\right]$$

in which the matrix L is the polynomial part of $\lambda^{-1}\left(\Gamma\lambda^{-1}+M+L\lambda\right)$. As explained in Appendix 2, these are the Hamilton equations for the functions H' and K respectively (recall from 1.1 that $H'=2H+(1-1/m)K^2$): here $\mathfrak{g}=\mathfrak{so}(3)[\lambda,\lambda^{-1}]$ is decomposed as the sum of the subalgebras of polynomials in λ and λ^{-1} and the invariant functions are $\operatorname{tr}(A^2)/2$ and $\lambda^{-1}\operatorname{tr}(A^2)/2$ respectively. Our two matrices $M+L\lambda$ and L are the respective gradients of these two functions at the point (Γ,L).

2.2. The spectral curve

As for any Lax equation, that of Proposition 2.1.1 describes isospectral variations of a matrix. The coefficients of the characteristic polynomial are thus first integrals. Moreover, because of the remark above, and according to the AKS theorem (Appendix 2), the functions H' and K commute on the orbit $\mathcal{O}_c$ – but we have already checked in I.2.1 that H and K commute, and the two statements are equivalent since H' is a linear combination of H and K^2.

The equation of the spectral curve is

$$\det\left(\Gamma\lambda^{-1}+M+L\lambda-\mu I\right)=0$$

where the vector $\Gamma\lambda^{-1}+M+L\lambda$ is considered as a skew-symmetric matrix, so that 0 is an eigenvalue and the characteristic polynomial can be written $\mu\left(\mu^2+Q(\lambda)\right)$ with

$$\begin{aligned}Q(\lambda)&=\left\|\Gamma\lambda^{-1}+M+L\lambda\right\|^2=-\tfrac{1}{2}\operatorname{tr}\left(\Gamma\lambda^{-1}+M+L\lambda\right)^2\\&=\lambda^{-2}+2M\cdot\Gamma\lambda^{-1}+\|M\|^2+2\Gamma\cdot L+2M\cdot L\lambda+\lambda^2.\end{aligned}$$

The shift in the exponents of λ that we have used shows that the two momenta $c = M \cdot \Gamma$ and $K = M \cdot L$ play analogous roles. If our point (M, Γ) belongs to the level (h', k) of (H', K) on the orbit $\mathcal{O}_c$, then

$$Q(\lambda) = \lambda^{-2} + 2c\lambda^{-1} + h' + 2k\lambda + \lambda^2.$$

Of course, the coefficients are the original first integrals (and we get nothing new!). The equation of the spectral curve indexed by (h', k) (or the corresponding level (h, k)) is

$$\mu\left(\mu^2 + Q(\lambda)\right) = 0.$$

It will be more convenient[4] to use, rather, the affine curve X_0,

$$\mu^2 + \lambda^{-2} + \lambda^2 + 2\left(c\lambda^{-1} + k\lambda\right) + h' = 0. \tag{6}$$

But it goes without saying that what we really want to use is the curve completed and normalised at $\lambda = 0$ and $\lambda = 0$ (see A.4.1). The most agreeable thing to do here is to consider that (6) is an equation in $\mathbf{P}^1 \times \mathbf{P}^1$: the two variables λ and μ are completed independently. At $\lambda = \infty$, for instance, the curve has two branches that are pure imaginary and conjugated, so that the smooth completion has two (conjugated imaginary) points a_- and b_- over $\lambda = \infty$ ($\lambda = \infty$, $\mu = \pm i\infty$). The situation at $\lambda = 0$ is quite similar. There the two points will be called a_+ and b_+. The curve smoothly completed in this way is a double cover of $\mathbf{P}^1$ branched at the four roots of the polynomial $\lambda^2 Q(\lambda)$; thus it is a genus-1 curve, which we will call the spectral curve and denote by X.

We already know that there should be an elliptic curve in this problem (see 1.1). The relations between the two curves C and X will be discussed in 2.4.

2.3. The eigenvector mapping

The spectral curve tells us something (actually everything) about the (non-zero) eigenvalues of the Lax matrix A_λ. Let us now consider its eigenvectors. If $(\lambda, \mu) \in X$ then, in general, A_λ has an eigenline for the eigenvalue μ. This statement is wrong at exactly four points, those where $\mu = 0$. There is nevertheless a limit line at these points, more precisely:

2.3.1 PROPOSITION. *If the curve X is smooth, there exists a complex line bundle of degree -4 over X, the fibre of which at a point (λ, μ) ($\mu \neq 0$) is the eigensubspace of the Lax matrix A_λ for the eigenvalue μ.*

This is a direct application of the general result A.3.1.1. The assertion on the degree can be obtained as a consequence of the Grothendieck-Riemann-Roch theorem (see A.4.1), as in Appendix 3 but with a slight modification since the sum of the two eigensubspaces corresponding to the two different values of μ over a given λ is not the whole space $\mathbf{C}^3$: we have discarded the kernel of A_λ. In any case, we are going to find an explicit divisor[5] representing this bundle, so that the assertion on the degree will be obvious.

---

[4]This way we are ignoring a genus-0 component and its intersection points with X_0. These would give rise to non-compact summands in the (generalised) Jacobian of the whole spectral curve. This is not very important but could nevertheless suggest another approach as explained by Adler & van Moerbeke in [2]. Also, the process is not completely innocent (see 2.4).

[5]For divisors, line bundles and Picard groups, see Appendix 4 and the references therein.

2.3.2 PROPOSITION. *Let a_- and b_- be the two points of X over $\lambda = \infty \in \mathbf{P}^1$. Let $R_\pm$ be the two points of X defined by*

$$\lambda(R_\pm) = -\frac{\gamma_1 \mp i\gamma_2}{u \mp iv}, \quad \mu(R_\pm) = \pm i\left(\gamma_3 \lambda(R_\pm)^{-1} + w + \lambda(R_\pm)\right).$$

There exists a non-vanishing section of the eigenvector bundle, the divisor of poles of which is $R_+ + R_- + a_- + b_-$.

This statement is precisely what we need to end the proof of 2.3.1. Let us prove it.

Proof. Consider the skew-symmetric matrix

$$A = \begin{pmatrix} 0 & -z & y \\ z & 0 & -x \\ -y & x & 0 \end{pmatrix}.$$

If μ is an eigenvalue of A, so that $\mu^2 + x^2 + y^2 + z^2 = 0$, then

$$V = \begin{pmatrix} \dfrac{-xz - \mu y}{x^2 + y^2} \\ \\ \dfrac{-yz + \mu x}{x^2 + y^2} \\ \\ 1 \end{pmatrix}$$

is an eigenvector of A for μ, that is obviously non-vanishing, taking the third coordinate into account. If A is the matrix $\Gamma\lambda^{-1} + M + L\lambda$, this gives us a non-vanishing section $V(\lambda, \mu)$ of the eigenvector bundle. The only thing to do now is to look at the poles of $V(\lambda, \mu)$, that is, at the points $(\lambda, \mu) \in X$ for which one or the other of the first two coordinates goes to infinity.

Now it is easily checked that

$$\left(f_+ = \frac{\mu + iz}{x - iy}, \quad f_- = \frac{\mu - iz}{x + iy}\right)$$

is a basis of the complex vector space generated by

$$\frac{-xz - \mu y}{x^2 + y^2} \quad \text{and} \quad \frac{-yz + \mu x}{x^2 + y^2}.$$

So we just need to find the poles of f_+ and f_- and thus, up to a change in sign of i, those of f_+. The denominator of f_+ vanishes for $x - iy = (\gamma_1 - i\gamma_2)\lambda^{-1} + (u - iv) = 0$, in other words for $\lambda = -(\gamma_1 - i\gamma_2)/(u - iv)$, with two possible values of μ; for one of these the numerator also vanishes. Remember that

$$(\mu + iz)(\mu - iz) + (x + iy)(x - iy) = 0.$$

The point obtained this way is the R_+ of our statement. It is a simple pole of f_+. Similarly, R_- is a simple pole of f_-.

The numerator $\mu + i(\gamma_3\lambda^{-1} + w + \lambda)$ goes to infinity for $\lambda = 0$ or $\lambda = \infty$. At $\lambda = 0$, the denominator also is infinite. Thus the points a_- and b_- are simple poles of f_+ and f_- and so of the first two coordinates of the eigenvector. □

Remark. Since $\mu^2 + x^2 + y^2 + z^2 = 0$, one has $f_+ f_- = -1$; thus the poles of f_+ are the zeros of f_- and, in particular, $R_+ + a_- \sim R_- + b_-$ (this is the *linear equivalence* relation, see Appendix 4).

An actual value (h', k) of the first integrals defines both a curve X and a level that can be indexed by (h', k) or by X itself. Let us call $\mathcal{T}_X$ such a level. What we have defined can be thought of as a mapping, the eigenvector mapping

$$f_X : \mathcal{T}_X \longrightarrow \operatorname{Pic}^4(X),$$

which associates with any point (Γ, M) of $\mathcal{T}_X$ the line bundle dual to that of eigenvectors, that is, the line bundle whose fibre at $(\lambda, \mu) \in X$ is the (dual line of the) eigenline of $\Gamma\lambda^{-1} + M + L\lambda$ with respect to the eigenvalue μ. We even now have an explicit representative of the class of the image.

The most classical use made of this mapping is the linearisation of the flows:

2.3.3 PROPOSITION. *The mapping f_X linearises the flows of H', K and H.*

This is a direct application of the general linearisation statement A.3.3.2. □

Another possible application is the determination of the regular values. It is based on Proposition A.3.2.3. The cocycles associated with H' and K by this proposition are those defined respectively by μ and $\mu\lambda^{-1}$. One checks quite easily that $\mu\lambda^{-1}$ is holomorphic in the neighbourhood of $\lambda = \infty$, so that it defines the zero class in the dimension-1 vector space $H^1(X; \mathcal{O}_X)$, the tangent space to $\operatorname{Pic}^4(X)$. This means that $X_K(\Gamma, M)$ belongs to the kernel of the tangent mapping $T_{(\Gamma,M)} f_X$. It actually generates it, as μ is certainly non-zero in this cohomology group. Thus the two vectors $X_{H'}$ and X_K are independent and we have proved:

2.3.4 PROPOSITION. *If the complex curve X is smooth, the point $(h, k) \in \mathbf{C}^2$ is a regular value of the momentum mapping (H, K) on the orbit $\mathcal{O}_c$.* □

This statement will be completed (by adding an "only if" part and a "real" part) in Proposition 2.5.1. Notice, however, that it is very easy to decide whether the curve is smooth: it is so exactly when the four roots of the polynomial $\lambda^2 Q(\lambda)$ above are distinct.

Remark. If one insists on not using the symmetries of the problem, then as an alternative it is possible to replace the Jacobian of X with a two-dimensional Jacobian, as done by Gavrilov & Zhivkov [33]. Notice that the highest-degree term in the Lax matrix A_λ is L, a *constant* matrix, so that one can choose an isomorphism of the fibres at a_- and b_- of the eigenvector bundle. The eigenvector mapping

$$f_X : \mathcal{T}_X \longrightarrow \operatorname{Pic}^4(X)$$

thus factorises through the set of isomorphism classes of line bundles $E \to X$ with an isomorphism $E_{a_-} \to E_{b_-}$. This is nothing other than the Jacobian of the singular curve X_s obtained from X by the identification of a_- and b_-, and this is an extension of $\operatorname{Pic}(X)$ by a multiplicative group $\mathbf{C}^\star$. I do not want to employ generalised Jacobians here: the Jacobian of X is simpler and will be sufficient for our purposes.

2.4. Further comments and properties

As it turns out, this "simple" example is actually not all that simple. There are several reasons for this.

Firstly, the flow of K is periodic, generating the group of rotations around the axis, so that the circle group $\mathcal{R}$ acts on the whole system and it might therefore seem clever to look at the reduced system obtained by quotienting everything by $\mathcal{R}$ (see 1.2). However, the Lax equation describes the unreduced system.

Moreover, the Lax equation is in $\mathfrak{so}(3)[\lambda, \lambda^{-1}]$ and nothing can prevent a 3×3 skew-symmetric matrix from having the eigenvalue 0. The spectral curve we are using is only one irreducible component of the curve defined by the characteristic polynomial. This looks quite innocent but actually raises some technical problems. For instance, the sum of the eigenspaces corresponding to a given λ is not the whole space $\mathbf{C}^3$. In more technical terms, the direct image $\lambda_\star L$ of the eigenvector bundle L on $\mathbf{P}^1$ is not the trivial plane bundle. Also, $h^0(L) = 2$ so that it is more difficult to reconstruct our space $\mathbf{C}^3$ starting from the vector bundle L (this is one of the points I have never understood in Verdier's paper [84]).

Remark. An easy way to make the spectral curve irreducible is to use $\mathfrak{su}(2)$ in place of $\mathfrak{so}(3)$ (we will do something analogous in Chapter III; see also Gavrilov & Zhivkov [33]).

The relation between the topological aspects (regular levels) and the curve X is not very clear, and for the same reasons as above. It should be noticed that the best papers[6] on the subject are either imprecise or wrong on this topic. It is not true that $\mathcal{T}_X/\mathcal{R}$ is the real part of X (which is empty), nor even that it is the real part of the Jacobian (which has two connected components although $\mathcal{T}_X/\mathcal{R}$ is connected). It is not true that f_X defines an isomorphism (in which category?) of $\mathcal{T}_X/\mathcal{R}$ onto its image. Here is a precise statement:

2.4.1 THEOREM. *Let $\mathcal{R}$ be the group of rotations around the axis of the top. If the complex curve X is smooth, the mapping $f_X : \mathcal{T}_X \to \operatorname{Pic}^4(X)$ defines a connected double covering of $\mathcal{T}_X/\mathcal{R}$ onto one of the two components of* $\operatorname{Pic}^4(X)_{\mathbf{R}}$.

Proof. The real structure S of $\operatorname{Pic}^d(X)$ is defined by complex conjugation on X: this defines an involution on divisors and, at the quotient level, an anti-holomorphic involution S on $\operatorname{Pic}^d(X)$ (see Appendix 4). For example, the divisor $a_- + b_-$ consists of the points $(\infty, \pm i\infty)$ in X; it is thus invariant by S, that is, real. As this divisor is constant, we take no risk in replacing $f_X : \mathcal{T}_X \to \operatorname{Pic}^4(X)$ by the mapping

$$\begin{array}{ccc} \mathcal{T}_X & \longrightarrow & \operatorname{Pic}^2(X) \\ (\Gamma, M) & \longmapsto & [R_+ + R_-], \end{array}$$

which will still be denoted f_X and considered as defined on the complex level $\mathcal{T}_X^{\mathbf{C}}$ (letting Γ and M take complex values, satisfying the same equations). Then f_X is a real mapping, in the sense that it preserves the real structures: since

$$R_\pm(\overline{\Gamma}, \overline{M}) = \overline{R_\mp(\Gamma, M)},$$

[6] Even the one of Audin & Silhol [15] and the French version *"Toupies"* of these notes!

it is true that $f_X(\overline{\Gamma}, \overline{M}) = Sf_X(\Gamma, M)$. In particular, f_X will send our real level $\mathcal{T}_X$ into the real part (the fixed points of S) of $\mathrm{Pic}^2(X)_{\mathbf{R}}$. Remember now that, for a genus-1 real curve X without real points, $\mathrm{Pic}^{2d}(X)$ is a real genus-1 curve, isomorphic to X over $\mathbf{C}$, but whose real part has two connected components (see Corollary A.4.5.2). Notice now that the image of f_X is contained in one[7] of the components: let us define a mapping

$$\begin{array}{rcl} g : X & \longrightarrow & \mathrm{Pic}^2(X) \\ P & \longmapsto & [P + \overline{P}]. \end{array}$$

Since the complex curve X is connected, and g is a continuous mapping, its image must be connected. Since it is included in the real part, it must be one of the two components, say C, of $\mathrm{Pic}^2(X)_{\mathbf{R}}$. But, if $(\Gamma, M) \in \mathcal{T}_X^{\mathbf{R}}$, then $f_X(\Gamma, M) = g(R_+(\Gamma, M))$ so that the image of the real level is included in the image of g: one component only can be reached.

Let us now fix a point $D = [R_+ + R_-]$ in the image. We look for all the $(\Gamma, M) \in \mathcal{T}_X$ such that $f_X(\Gamma, M) = D$.

Notice first that R_+ determines (Γ, M) up to the flow of K. Knowing

$$\lambda = -\frac{\gamma_1 - i\gamma_2}{u - iv},$$

$\mu = i(\gamma_3\lambda^{-1} + K + \lambda)$, so that γ_3 can be deduced, then both $|\gamma_1 - i\gamma_2|^2 = 1 - \gamma_3^2$ and $|u - iv|^2$ are known, thanks to H', and $\mathrm{Re}\,[(\gamma_1 - i\gamma_2)(u + iv)]$, which equals c. So, $\gamma_1 - i\gamma_2$ and $u - iv$ are known up to multiplication by (the same) $e^{i\theta}$, and eventually we deduce the class of (Γ, M) in $\mathcal{T}_X/\mathcal{R}$.

To prove that the covering has order 2, it is enough to find one point in the image whose pre-image consists of two points of $\mathcal{T}_X/\mathcal{R}$. Consider the class D of $a_- + b_-$ in $\mathrm{Pic}^2(X)$, so that, for the Abel-Jacobi mapping u,

$$u^{-1}(D) = \{P + \tau P \mid P \in X\} \subset X^{(2)}$$

where τ is, of course, the elliptic involution $(\lambda, \mu) \mapsto (\lambda, -\mu)$ of X and $X^{(2)}$ is its symmetric square (see Appendix 4). It is easily checked (and is left as an exercise for the reader) that the inverse image of D in $\mathcal{T}_X/\mathcal{R}$ consists of the (Γ, M) sent to $R_+ + R_-$, with

$$R_+ = \left(-\frac{1}{\alpha}, i\left(-\gamma_3\alpha + k - \frac{1}{\alpha}\right)\right) \quad \text{and} \quad R_- = \tau R_+,$$

where γ_3 is one of the roots, x_3 or x_2, of the polynomial f of equation (3) and where

$$\alpha = \frac{c - k\gamma_3}{1 - \gamma_3^2}.$$

The two values of R_+ determined this way correspond to two classes of (Γ, M) modulo $\mathcal{R}$ sent to the class $R_+ + R_-$. Thus our covering indeed has order 2. Actually $\mathcal{T}_X/\mathcal{R}$ is endowed with an involution defined according to the remark following the proof of Proposition 2.3.2, so that

$$\begin{cases} R_+ + R_- & \sim \; D \\ R_+ - R_- & \sim \; a_- - b_- \end{cases}$$

and the point R_+ is well determined up to an element of order 2. There are three non-zero elements of order 2 in $\mathrm{Pic}^0(X)$, but only one of them can send a point $R_+(\Gamma, M)$ to a point of the same form $R_+(\Gamma', M')$, as it is clear in the example above. This defines an involution σ on $\mathcal{T}_X/\mathcal{R}$.

[7] I don't want to use, at this point, the fact that we know that the level $\mathcal{T}_X$ is connected.

That the covering is non-trivial is a consequence of the fact that the levels $\mathcal{T}_X$ are connected, which we have already proved in 1.2 and 1.3 and will re-prove in 2.6. Nevertheless, I would have preferred to use the eigenvector mapping to prove that fact. With coverings, though, this would be rather delicate. □

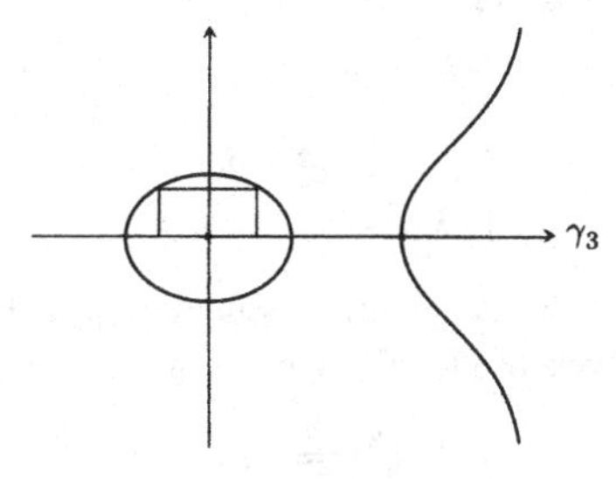

Figure 8

Remarks

- The transformation induced by the involution σ_2 on γ_3 is shown in Figure 8 (the curve $y^2 = f(x)$ is an avatar of X with two real components, that is, an avatar of $\mathrm{Pic}^2(X)$, see Appendix 4); the order-2 elements are the points of coordinates $(x_i, 0)$, the one that allows us to go from R_+ to R_- being $(x_1, 0)$).

- The proof shows that the complexified map

$$f_X^{\mathbf{C}} : \mathcal{T}_X^{\mathbf{C}}/\mathcal{R}^{\mathbf{C}} \longrightarrow \mathrm{Pic}^2(X)$$

  has degree 4 (onto its image).

2.5. Regular and critical values

The spectral curve X describes the (non-zero) eigenvalues of the skew-symmetric matrix $A_\lambda = \Gamma\lambda^{-1} + M + L\lambda$. The complex curve $X_{\mathbf{C}}$ is smooth exactly when the polynomial $\lambda^2 Q(\lambda)$ has no multiple root. We give now a more complete statement than 2.3.4 (and a proof) on the regular values of (H', K). Notice in particular that X has no real point in this case: $Q(\lambda) = \|\Gamma\lambda^{-1} + M + L\lambda\|^2 > 0$, $X_{\mathbf{R}} = \varnothing$.

2.5.1 PROPOSITION. *The point $(h, k) \in \mathbf{R}^2$ is a regular value of the momentum mapping (H, K) on the orbit $\mathcal{O}_c$ if and only if the corresponding complex curve $X_{\mathbf{C}}$ is smooth.*

Proof. Let us look first at the curve. The complex curve $X_{\mathbf{C}}$ is smooth at $\lambda = 0$ and $\lambda = \infty$ by definition, so that it is singular if and only if the polynomial $\lambda^2 Q$ has a multiple root. This is a degree-4 real polynomial, and it is nonnegative for real values of λ, so that if it has multiple roots, they can be either one (or two) real double root(s) or two double roots that are complex conjugates. Moreover, it has a real double root if and only if it has a real root, that is if and only if

$$\exists \lambda \in \mathbf{R} \mid \Gamma\lambda^{-1} + M + L\lambda = 0,$$

and it has two conjugate double roots if and only if it is the square of an irreducible real polynomial, that is if and only if

$$\begin{aligned}\lambda^2 Q &= 1 + 2c\lambda + h'\lambda^2 + 2k\lambda^3 + \lambda^4 \\ &= (1 + c\lambda + \lambda^2)^2.\end{aligned}$$

This can only happen if

$$\begin{cases} h' &= c^2 + 2 \\ k &= c, \end{cases}$$

with $c^2 - 4 < 0$ in order that the two double roots be non-real.

Let us look now at the levels of (H, K). We are on the orbit $\mathcal{O}_c$,

$$\begin{cases} \|\Gamma\|^2 &= 1 \\ \Gamma \cdot M &= c, \end{cases}$$

and we wish to know when the functions

$$\begin{cases} H(\Gamma + \varepsilon M) &= \frac{1}{2}\Omega \cdot M + \Gamma \cdot L \\ K(\Gamma + \varepsilon M) &= L \cdot M \end{cases}$$

are independent. We have already computed the gradients (see I.2.1):

$$\begin{aligned}\nabla_{\Gamma+\varepsilon M} H &= \Omega + \varepsilon L \\ \nabla_{\Gamma+\varepsilon M} K &= L.\end{aligned}$$

Using the same invariant bilinear form, the gradients of the two functions that define the orbit $\mathcal{O}_c$ are obviously $\varepsilon\Gamma$and $\Gamma + \varepsilon M$ respectively. The question of the regular levels is that of the rank of the 6×4 matrix

$$\begin{pmatrix} 0 & \Gamma & \Omega & L \\ \Gamma & M & L & 0 \end{pmatrix}$$

and this is not very difficult to determine. Remember that $M - \Omega$ is collinear with L (or use H' in place of H) so that this rank is that of

$$\begin{pmatrix} 0 & \Gamma & M & L \\ \Gamma & M & L & 0 \end{pmatrix}.$$

To say that the rank is not maximal is to say that the three vectors Γ, M, L of $\mathbf{R}^3$ are dependent.

1. If the vectors L, Γ, M are collinear – the axis of the top is vertical and so is the angular momentum – the 6×4 matrix can be written

$$\begin{pmatrix} 0 & \Gamma & s\Gamma & t\Gamma \\ \Gamma & s\Gamma & t\Gamma & 0 \end{pmatrix}$$

for real numbers s and t. At such a point, the momentum mapping (H, K) has rank 0 and takes the value

$$h = \frac{c^2}{2m} + \eta, \quad k = c\eta \quad \text{for } \eta = \pm 1$$

or

$$\begin{cases} h' &= c^2 + 2\eta \\ k &= c\eta \end{cases}$$

according to the previous formulae. For $\eta = 1$ and $c < 2$ this is the point where $\lambda^2 Q$ has two non-real conjugate double roots.

2. If the three vectors generate a dimension-2 subspace of $\mathbf{R}^3$, we have two linear relations

$$\beta\Gamma + \gamma M + \delta L = 0$$

and

$$\alpha\Gamma + \beta M + \gamma L = 0,$$

expressing that the rank is < 4. They cannot be independent. Dividing the first one by γ and the second one by β brings them to the form

$$\begin{cases} a\Gamma + M + \lambda L = 0 \\ b\Gamma + M + a^{-1} L = 0 \end{cases}$$

so that $a = \lambda^{-1}$ and there exists a real number λ such that $\Gamma\lambda^{-1} + M + L\lambda = 0$. In other words, $\lambda^2 Q$ has a real (double) root and the rank of the momentum mapping is 1. □

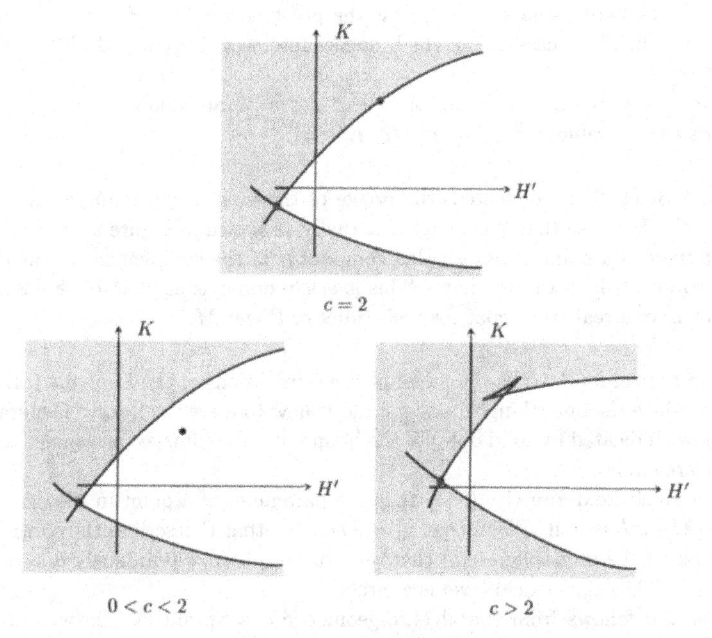

Figure 9: The momentum mapping (H', K) for the symmetric top.

Figure 9 shows the image of the momentum mapping (its regular and its critical values) for $c \geq 0$ (one easily deduces the $c \leq 0$ cases: changing M to $-M$ changes c to $-c$ and

K to $-K$ and leaves $\|\Gamma\|^2$, H' and H unchanged). The two solid dots in each diagram show the values of (H', K) for which the polynomial has two double roots, and the curve corresponds to a real double root. Notice that in the lower left-hand diagram, one of the solid dots is in the interior of the image, not on the critical curve: this corresponds here to two non-real double roots – and also to a singular $X_{\mathbf{C}}$ with smooth (actually empty) real part. Indeed, Proposition 2.5.1 relates the smoothness of the *real* levels to that of the *complex* curve.

2.6. Singularities

Let us now explain how the diagrams in Figure 9 were obtained, and make some comments on the image, the singular values and critical levels.

The discriminant. Of course, these diagrams show the sections of the discriminant of all the polynomials of degree 4 (the swallow tail). The discriminant curve is parametrised by the double root t: we simply write that $\lambda^2 Q(\lambda) = (\lambda - t)^2(\lambda^2 + a\lambda + b)$ to get

$$\begin{cases} H' &= t^2 - 4ct^{-1} - 3t^{-2} \\ K &= -t + ct^{-2} + t^{-3}. \end{cases}$$

It has two branches, for $t < 0$ and for $t > 0$. The change of c to $-c$ exchanges the two branches, so that we see again that it is sufficient to look at the $c \geq 0$ case. The two branches intersect when $\lambda^2 Q$ has two real double roots, that is, when $t^2 - ct + 1 = 0$ (in this case, the two branches intersect at the point $(H', K) = (c^2 - 2, -c)$) and when $t^2 + ct + 1 = 0$, in which case one of the branches intersects itself at $(H', K) = (c^2 + 2, c)$ if $c^2 - 4 \geq 0$.

When $c^2 - 4 < 0$, the two roots of $t^2 + ct + 1 = 0$ are double roots of $\lambda^2 Q$; this corresponds to real values $(c^2 + 2, c)$ of (H', K).

The image. Since $\mathcal{O}_c$ is connected, the image of the momentum mapping will be connected as well. To prove that this image is actually as shown in Figure 9, we need only to check that there is a point in any shaded component of the complement of the discriminant that cannot belong to the image. This is easily done, using that $H' \geq -2$ and that $\lambda^2 Q$ cannot have a real triple root for real values of Γ and M.

The generic critical levels. The problem of the bifurcations of the Liouville tori is quite simple here, since the only thing to understand is how to leave the image. General points (i.e. not those indicated by solid dots) of the boundary of the image correspond to critical levels that are circles.

This can be checked directly. If $t \in \mathbf{R}$ is the parameter of a point in the discriminant curve, $\Gamma + tM + t^2 L = 0$; it follows that $\gamma_3 = -tK - t^2$, that Γ describes the corresponding parallel circle of the unit sphere and that $\gamma_1 + tu = \gamma_2 + tv = 0$ uniquely determines M in terms of Γ. Thus the critical level is a circle.

But this also follows from the algebro-geometric description in 1.2: we still have a circle-bundle over an oval, but the oval has become a single point. In terms of Figure 7: the spherical zone has narrowed to become a circle.

Notice that the direct proof gives once again that these levels – and hence all non-empty levels – are connected.

The centre-centre and focus-focus points. The solid dots in Figure 9 correspond to levels that contain a point at which the momentum mapping

$$(H', K) : \mathcal{O}_c \longrightarrow \mathbf{R}^2$$

has rank 0. Either directly or using the algebro-geometric descriptions at hand, it is easily seen that the levels of the solid dots on the boundary (that is, the corners of the boundary, where $(h', k) = (c^2 + 2\eta, \eta c)$ for $\eta = -1$ and 1, and $c^2 \geq 4$) consist of single critical points. It seems that we ought to call such points *centre-centre points*. The situation here is very similar to what happens in the neighbourhood of a fixed point for a Hamiltonian T^2-action on a symplectic 4-manifold.

This is quite different at the solid dot inside the image ($(h', k) = (c^2 + 2, c)$ for $c^2 < 4$). The pre-image of such a point is a sphere with two points identified (this is the critical point on this level). Such points are called *focus-focus points*.

More on the focus-focus point. The singular point inside the image is something very surprising for a singularity of differentiable mapping, but it is nevertheless very common in integrable systems with two degrees of freedom (see also Figure 17 in Chapter IV). It appears in the classification of Abelian subalgebras of the Lie algebra of the symplectic group of $\mathbf{R}^4$ or, which amounts to the same thing, in that of Abelian subalgebras of the homogeneous quadratic polynomials on $\mathbf{R}^4$ with the usual Poisson bracket (this is a theorem of Williamson [88], explained by Arnold in Appendix 6 of [9], and used e.g. by Lerman & Umanskii [58] and Desolneux-Moulis [24]). It is a general feature of focus-focus points that their inverse image has to be a chain of spheres (here just one sphere), each intersecting the following at a point, as was shown by Nguyen [67].

The "disappearance" of the focus-focus point (its transformation into a centre-centre point when c reaches the value 2) is sometimes called a *Hamiltonian Hopf bifurcation* (see, for this example, especially Cushman & van der Meer [23]).

The choice of the momentum mapping and convexity. Although this is not very common, the picture of the regular values here is not something new; it can be found in at least, besides my joint paper with Silhol [15], Cushman & Knörrer [22] and in Fomenko [32] (I have not tried to give a complete list of references for this point; it is certain, however, that the question was not of particular interest for the authors of the books and papers I have already mentioned). However, the pictures obtained from the two last-mentioned works are somewhat different, although non-contradictory, from those in Figure 9. The reason is that we have chosen here to draw the image of the momentum mapping (H', K) on a given orbit $\mathcal{O}_c$. Another possible choice is to use (H, K), as in [32], and a third is to look at the three functions (c, H, K) together, as in [22].

I have explained in the Introduction that there is no canonical choice for the momentum mapping: the only thing that is given intrinsically is an Abelian subalgebra of the Lie (Poisson) algebra of all functions: (H', K) is not a better choice than (H, K), because there are no "better" choices. However, the algebra of all first integrals contains a subalgebra that is intrinsically defined, that of the Casimir functions, so that c does not play the same role here as H and K (one should actually add $\|\Gamma\|^2$ to the picture).

The choice of (H', K) was suggested here by the specific form of the Lax matrix. Once the Lax matrix is given, it is obvious (see the differential equations at the end of 2.1) that H' should be used, but, if we think of the origin of the problem, then one should obviously use H, the total energy of the system.

Now it is a value of H' that appears in the equation of the spectral curve and this is why it was so easy to get the image; we just have had to look at the multiple roots of a polynomial. This is also why there is some kind of convexity in the picture: our boundary is a linear section of the discriminant of all the degree-4 polynomials, so that it may be described as the envelope of a smooth family of lines, and it cannot have points of inflexion. This is a general feature of momentum mappings given by the coefficients of a polynomial (see the figures in chapters III and IV and in the author's papers on Moser systems [13, 11]). Of course nothing prevents the critical curve for the momentum mapping (H, K) from having points of inflexion, and it actually does have one (see the figures in the above-mentioned papers).

III

The Kowalevski top

This chapter is devoted to the Kowalevski top. I will first try to give a flavour of the results and methods of Sophie Kowalevski's original paper [55]. I will then explain how the eigenvectors of the Lax matrices allow us to understand the topology ... once the Lax matrices are known. These results are based on the work [18] of Bobenko, Reyman & Semenov-Tian-Shanski, and were obtained jointly with Robert Silhol [15]. The latter paper contains complete proofs and detailed computations that I will not reproduce here: I send the interested readers (if any) to the original paper for the missing details. In the last section of this chapter, I will explain briefly the remarkable construction of Bobenko, Reyman & Semenov-Tian-Shansky [18] leading to the Lax pair used up to now. This will lead us to two examples intended to illustrate the limits of the methods used in this book.

1. Kowalevski's method

I use the notation of I.2.1, except that the orbits will be indexed by $2l = c$, as is usual in the classical literature.

1.1. A few algebraic reductions

Let me recall the method used by Kowalevski in [55]. The second first integral K being

$$K = \left|(p+iq)^2 + (\gamma_1 + i\gamma_2)\right|^2,$$

it is natural to put $x = p + iq$ and

$$\begin{aligned} \xi &= p^2 - q^2 + \gamma_1 + i(2pq + \gamma_2) \\ &= x^2 + \gamma_1 + i\gamma_2, \end{aligned}$$

and this is indeed[1] what she does. It is convenient to use also $y = p - iq$ and $\eta = p^2 - q^2 + \gamma_1 - i(2pq + \gamma_2)$ and to consider x, y, ξ and η as independent complex variables (they are actually independent once the equations are complexified, allowing M and Γ

[1] I have used the non-lining numerals as in *Acta Math.*, for those formulae that come directly from [55].

to live in $\mathbf{C}^3$ rather than in $\mathbf{R}^3$). This way, the following four equations (the first two describe the orbit $\mathcal{O}_{2l}$, the next two the integrals H and K),

$$\begin{cases} 1 &= \|\Gamma\|^2 \\ 2l &= \Gamma \cdot M \\ H &= \frac{1}{2}\Omega \cdot M + \Gamma \cdot L \\ K &= (p^2 - q^2 + \gamma_1)^2 + (2pq + \gamma_2)^2, \end{cases}$$

become

$$\begin{cases} \gamma_3^2 &= 1 - K + y^2\xi + x^2\eta - x^2y^2 \\ r\gamma_3 &= 2l - y\xi - x\eta + xy(x+y) \\ r^2 &= 2H + \xi + \eta - (x+y)^2 \\ \xi\eta &= K. \end{cases}$$

She eliminates r and γ_3 between the first three of the latter group of equations, using the polynomials

$$R(x) = -x^4 + 2Hx^2 + 4lx + 1 - K$$

and

$$R_1(x,y) = -2Hx^2y^2 - 4lxy(x+y) - (1-K)(x+y)^2 + 2H(1-K) - 4l^2,$$

which allow her to write

$$\xi R(y) + \eta R(x) + R_1(x,y) + K(x-y)^2 = 0.$$

She also remarks that

$$(1) \qquad -4\left\{\frac{dx}{dt}\right\}^2 = R(x) + (x-y)^2\,\xi,$$

and thus decides to consider the genus-1 curve $\mathcal{E}$ of the equation $u^2 = -R(x)$. Its Jacobian[2] is the (isomorphic) curve $\mathcal{E}'$ whose equation is $t^2 = S(s)$, with

$$S(s) = 4s^3 - \left(\frac{H^2}{3} - (1-K)\right)s + \frac{H}{3}\left(\frac{H^2}{9} + 1 - K\right) - l^2.$$

The curve $\mathcal{E}$ is a principal homogeneous space under the action of the group $\mathcal{E}'$. The signs + and − that I shall use refer to this action.

Consider the holomorphic form $\omega = dx/u$. The automorphisms of $\mathcal{E}$ that change ω into $\pm\omega$ have the form $M \mapsto P \pm M$ for a point P in $\mathcal{E}'$ (see the explanations of Horozov & van Moerbeke [43] and Weil [86]). In other words, the points P of $\mathcal{E}'$ parametrise these automorphisms.

This can be expressed by formulae. We look for the automorphisms $(x,u) \mapsto (y,v)$ ($v^2 = R(y)$) such that $\dfrac{dx}{u} = \pm\dfrac{dy}{v}$, in other words such that

$$\left\{\frac{dy}{dx}\right\}^2 = \frac{R(y)}{R(x)}.$$

[2] The difference between a genus-1 curve and its Jacobian is the following: the Jacobian is a group and must thus have a distinguished element. A genus-1 curve of the equation $t^2 = S(s)$ for a degree-3 polynomial S is a group (in the most classical way possible, using secants), the point at infinity playing the role of the unit (see Appendix 4).

The solutions of this differential equation are given by the equation

$$\Phi_s(x,y) = 0$$

for a homogeneous symmetric polynomial Φ in x and y, having degree 2 in each variable, and also of degree 2 in the parameter s. After some (!) computations, one eventually gets the equation

$$4(x-y)^2\left(s-\frac{H}{6}\right)^2 - 4\left(s-\frac{H}{6}\right)\widetilde{R}(x,y) + R_1(x,y) = 0,$$

which appears at page 188 in [55] and in which

$$\widetilde{R}(x,y) = -x^2y^2 + 2Hxy + 2l(x+y) + 1 - K.$$

Its significance is the following: let $M = (x,u)$ be a given point of $\mathcal{E}$ (notice, by the way, that $-M = (x,-u)$). Fix also $P = (s,t) \in \mathcal{E}'$ (similarly $-P = (s,-t)$). The solutions of the equation (in y) $\Phi_s(x,y) = 0$ are $y(P+M)$ and $y(P-M)$.

On the other hand, if, M still being fixed, a second point $M' = (y,v) \in \mathcal{E}$ is fixed, the solutions in s of $\Phi_s(x,y) = 0$ are

$$s_1 = s(M+M') = s(-M'-M) \quad \text{and} \quad s_2 = s(M-M') = s(M'-M). \tag{2}$$

This is the famous "mysterious change of variables" of Kowalevski: she replaces x and y by s_1 and s_2. Note that, by definition, using (2), we have

$$\begin{cases} \dfrac{ds_1}{\sqrt{S(s_1)}} & = & \dfrac{dx}{\sqrt{R(x)}} + \dfrac{dy}{\sqrt{R(y)}} \\[2ex] \dfrac{ds_2}{\sqrt{S(s_2)}} & = & -\dfrac{dx}{\sqrt{R(x)}} + \dfrac{dy}{\sqrt{R(y)}}. \end{cases}$$

1.2. Linearisation of the flow of H

Kowalevski then uses equation (1) and its twin:

$$\begin{cases} -4\left\{\dfrac{dx}{dt}\right\}^2 & = & R(x) + (x-y)^2\xi \\[2ex] -4\left\{\dfrac{dy}{dt}\right\}^2 & = & R(y) + (y-x)^2\eta \end{cases}$$

to get

$$-4\left\{\frac{1}{\sqrt{S(s_1)}}\frac{ds_1}{dt}\right\}^2 = 4\frac{(x-y)^4}{R(x)R(y)}\left(s_1 - \frac{H}{6} - \frac{\sqrt{K}}{2}\right)\left(s_1 - \frac{H}{6} + \frac{\sqrt{K}}{2}\right).$$

Writing

$$T(s) = -S(s)\left(s - \frac{H}{6} - \frac{\sqrt{K}}{2}\right)\left(s - \frac{H}{6} + \frac{\sqrt{K}}{2}\right)$$

so that

$$\frac{ds_1}{\sqrt{T(s_1)}} = \frac{dt}{s_1 - s_2}$$

and similarly

$$\frac{ds_2}{\sqrt{T(s_2)}} = \frac{dt}{s_2 - s_1},$$

she deduces the differential equations that s_1 and s_2 satisfy:

$$(3) \qquad \begin{cases} 0 = \dfrac{ds_1}{\sqrt{T(s_1)}} + \dfrac{ds_2}{\sqrt{T(s_2)}} \\[2ex] dt = \dfrac{s_1 ds_1}{\sqrt{T(s_1)}} + \dfrac{s_2 ds_2}{\sqrt{T(s_2)}}. \end{cases}$$

She thus considers the hyperelliptic curve X of equation $y^2 = T(s)$.

1.2.1 PROPOSITION. *The flow of H is linearised on the Jacobian of X.*

Proof. The first thing to do is to give a precise meaning to a formulation that is both classical and non-understandable. The curve X has genus 2 and the change of variables has given us the differential system (3), which I prefer to write as

$$(4) \qquad \begin{cases} 0 = \dfrac{ds_1}{y_1} + \dfrac{ds_2}{y_2} \\[2ex] dt = \dfrac{s_1 ds_1}{y_1} + \dfrac{s_2 ds_2}{y_2} \end{cases}$$

to understand it as a differential system on the symmetric product $X^{(2)}$. Consider the Abel-Jacobi mapping

$$\begin{array}{ccc} X^{(2)} & \xrightarrow{u} & \mathrm{Jac}(X) \\ P_1 + P_2 & \longmapsto & \left(\omega \mapsto \int_P^{P_1} \omega + \int_P^{P_2} \omega\right). \end{array}$$

Let

$$\begin{array}{ccc} \mathbf{R} & \xrightarrow{\gamma} & X^{(2)} \\ t & \longmapsto & ((s_1, y_1), (s_2, y_2)) \end{array}$$

be a solution of (4). The precise statement we want to prove is that $u \circ \gamma : \mathbf{R} \to \mathrm{Jac}(X)$ is (the image of) a line (of $\mathbf{C}^2$) with linear parametrisation, and that all these lines are parallel.

Now, the Jacobian $\mathrm{Jac}(X)$ is the quotient of the dual $H^0(\Omega_X^1)^*$ of the vector space of holomorphic forms on X and the period lattice Λ. In particular, any holomorphic form α

on X can be considered as a 1-form on Jac(X): thus $\alpha \in H^0(\Omega_X^1)$ is a (constant) 1-form on $H^0(\Omega_X^1)^\star$, that is, after going to the quotient, a 1-form on Jac(X). Here

$$\left(\frac{ds}{y}, \frac{sds}{y}\right)$$

is a basis of $H^0(\Omega_X^1)$. Call (x_1, x_2) the coordinates in the dual basis. Write $u \circ \gamma(t) = (x_1(t), x_2(t))$ so that system (4) can be read on Jac(X) as

$$\begin{cases} \dfrac{dx_1}{dt} &= 0 \\ \dfrac{dx_2}{dt} &= 1 \end{cases}$$

so that, of course, the flow lines are (images of) straight lines, with linear parametrisation. □

Remark. This statement can be used to express the solutions in terms of ϑ-functions.

The curve X. To simplify the formulae, let us put

$$z = 2\left(s + \frac{H}{3}\right),$$

so that

$$2S(s) = \varphi(z) = z^3 - 2Hz^2 + \left(H^2 + 1 - K\right)z - 2l^2$$

and

$$\left(s - \frac{H}{6} - \frac{\sqrt{K}}{2}\right)\left(s - \frac{H}{6} + \frac{\sqrt{K}}{2}\right) = \frac{1}{2}\left(z - H - \sqrt{K}\right)\left(z - H + \sqrt{K}\right)$$

The equation of the curve X becomes

$$y^2 = -2\varphi(z)\left[(z-H)^2 - K\right].$$

This does not prevent X from being a (hyperelliptic) genus-2 curve, smooth if and only if the polynomial

$$\Phi(z) = -2\varphi(z)\left[(z-H)^2 - K\right]$$

has only simple roots.

The discriminant of the family of curves X describes the values of H and K for which Φ has a multiple root. That φ has a multiple root gives a cusp curve, that $\left[(z-H)^2 - K\right]$ has a double root gives a line ($K = 0$), and that these two polynomials have a common root gives a parabola, tangent to both the cusp curve and the H-axis, which is the real part of this discriminant in the KH-plane[3] (see Figure 10).

[3] I have followed Kharlamov [47] in whose paper these diagrams first appeared and who used non-alphabetic order for the coordinates H and K.

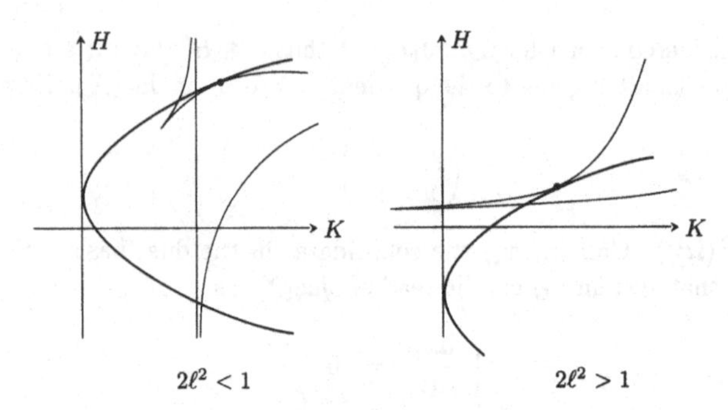

Figure 10: The discriminant in the Kowalevski case.

Remark. The number of components of the real part $X_{\mathbf{R}}$ of the curve X depends only on the number of real roots of Φ, that is, of φ (as the degree-2 factor always has two real roots). Thus, it does not depend on the position of the point (K, H) under consideration with respect to the parabola. The same is true for the number of components[4] of the real part $\text{Jac}(X)_{\mathbf{R}}$ of the Jacobian. We will see, however, that the number of Liouville tori in a given level can change from one side of the parabola to the other. What this remark intends to show is the following: the Kowalevski change of variables allows us to linearise the flow and to write the solutions of the Hamiltonian system (she has written the solutions quite explicitly in her paper) but they do not give directly the number of Liouville tori. The method "Lax pair/spectral curve/eigenvector mapping" has the advantage of giving a *mapping* from the Liouville tori to some Abelian varieties, which linearises the flows *and* gives information on the topology (as we shall be showing in this book). This is probably why we do not know any *real* Lax pair[5] for the Kowalevski top that gives the curve X as spectral curve.

However, using the Kowalevski change of variables and a large amount of computation, it is certainly possible to get information on the topology of the regular levels and their bifurcations. This was done by Kharlamov in [47]. His work was based on the "real motions" previously investigated by Appelrot [8]. This latter author had determined the image of the projection of Ω (or $\frac{1}{2}M$) on the equatorial plane of the body, i.e. he had determined which values of (p, q) occur. His investigation used the functions s_1 and s_2 (the ones used for the linearisation statement above) and their possible positions with respect to the real roots of the polynomial T. The first point is that on the one hand

$$s_2 < H - \sqrt{K} < s_1 < H + \sqrt{K}$$

and, on the other hand, that if e_1, e_2, e_3 (resp. e_1) are the real roots (resp. root) of φ, then

$$s_2 < e_3 < e_2 < e_1 < s_1 \quad (\text{resp. } s_2 < e_1 < s_1).$$

What Appelrot then did was to draw the level curves for the two functions s_1 and s_2 in the equatorial plane (pq-plane). The regions corresponding to real motions are those

[4]It is a classical result of Weichold and Klein that the number of connected components of $\text{Jac}(X)_{\mathbf{R}}$ depends only on that of $X_{\mathbf{R}}$, see Appendix 4.

[5]There exist Lax pairs for which X is the spectral curve (those constructed by Haine & Horozov [40] or Adler & van Moerbeke [4]), but they have non-real coefficients.

allowed by these inequalities. It is not very hard to imagine that the position of the point (K, H) under consideration with respect to the discriminant can be important: it determines the number of real roots of φ *and* their position with respect to $H \pm \sqrt{K}$.

Figure 11 comes from Appelrot's paper [8] and shows the shape of the region obtained according to which zone of Figure 12 (K, H) belongs (the zones are labelled respectively region 1, region 2, regions 3 and 4, region 4′, region 5).

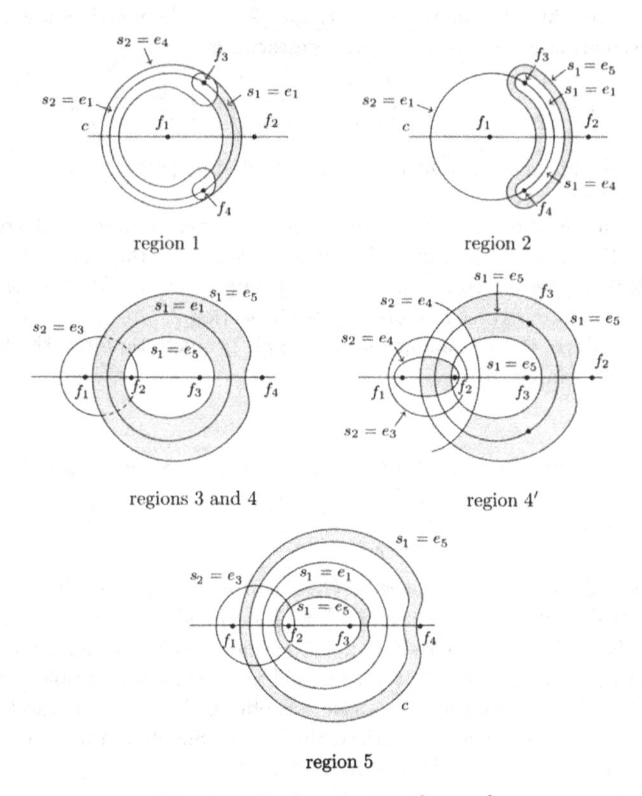

Figure 11: Real motions in the *pq*-plane.

What Kharlamov [47] later did was to compute of how many points (Γ, M) a given point (p, q) was the image. It is probably then possible to deduce the number of Liouville tori corresponding to any component of the complement of the discriminant.

It is even possible to investigate their bifurcations with the help of the bifurcations of the real-motion regions (which was also done by Appelrot). Although the proofs in Kharlamov's paper [47] are not complete (for the bifurcations, there is no proof at all), it must be noticed that this is (at least to my knowledge) the first paper in which the bifurcations of the Liouville tori were considered systematically and where all the possible local models for two degrees of freedom were shown.

2. Lax pair and spectral curves

2.1. The Lax pair

Bobenko, Reyman and Semenov-Tian-Shanski have constructed in [18] a very natural Lax pair

$$\dot{L_\lambda} = [L_\lambda, M_\lambda] .$$

The Lie algebra in which the matrices live is $\mathfrak{so}(3,2)$ and the matrices are given, in the usual block decomposition (see section 3 if necessary), by

$$L_\lambda = \begin{pmatrix} 0 & F \\ {}^tF & 0 \end{pmatrix} \lambda^{-1} + \begin{pmatrix} -M & 0 \\ 0 & PMP \end{pmatrix} + \begin{pmatrix} 0 & E \\ {}^tE & 0 \end{pmatrix} \lambda. \tag{5}$$

The matrix M_λ is, up to a constant factor, the polynomial part in $\operatorname{tr} L_\lambda^2$, and the Hamiltonian is $H = \frac{1}{8} \operatorname{Res}(\operatorname{tr} L_\lambda^2 \lambda^{-1} d\lambda)$.

In equation (5), E and F are matrices consisting of two columns that are vectors of $\mathbf{R}^3$; here $F = (\Gamma, 0)$ and E consists of the two first vectors in the canonical basis of $\mathbf{R}^3$, while PMP is the upper left corner of the skew-symmetric matrix M. A brief description of the construction of our three authors is given in section 3.

If $I_{3,2}$ is the matrix of the standard signature-$(3,2)$ quadratic form, the fact that L_λ belongs to $\mathfrak{so}(3,2)$ can be written

$$I_{3,2}{}^tL_\lambda = -L_\lambda I_{3,2}. \tag{6}$$

There is also a symmetry coming from the dependence in λ, the diagonal blocks being skew-symmetric:

$$L_{-\lambda} = -{}^tL_\lambda. \tag{7}$$

The equation of the spectral curve is $\det(L_\lambda - \mu \operatorname{Id}) = 0$. It has two involutions, τ_1 and τ_2, which reflect the relations above: $\tau_1(\lambda,\mu) = (-\lambda,\mu)$ and $\tau_2(\lambda,\mu) = (\lambda,-\mu)$.

From equation (6), and because $5 = 2+3$ is odd, 0 is always an eigenvalue of the matrix L_λ, so that the characteristic polynomial will define a reducible curve. I have already claimed that the Lax equation above was obtained in a very natural way in [18]. Now, the authors of this paper also noticed that it was possible to use the isomorphism $\mathfrak{so}(3,2) \cong \mathfrak{sp}(4,\mathbf{R})$ to replace L_λ by the matrix

$$L'(\lambda) = \left(\begin{array}{cc|cc} \gamma_1 & \gamma_2 & \gamma_3 & 0 \\ \gamma_2 & -\gamma_1 & 0 & -\gamma_3 \\ \hline \gamma_3 & 0 & -\gamma_1 & \gamma_2 \\ 0 & -\gamma_3 & \gamma_2 & \gamma_1 \end{array}\right) \lambda^{-1} + \left(\begin{array}{cc|cc} \multicolumn{2}{c|}{0} & -v & -u \\ & & u & -v \\ \hline v & -u & 0 & -2w \\ u & v & 2w & 0 \end{array}\right) + \left(\begin{array}{cc|cc} 0 & 0 & \multicolumn{2}{c}{0} \\ 0 & 0 & & \\ \hline \multicolumn{2}{c|}{0} & 2 & 0 \\ & & 0 & -2 \end{array}\right) \lambda, \tag{8}$$

which satisfies a Lax equation with M_λ replaced by a matrix $M'_\lambda \in \mathfrak{sp}(4)(\mathbf{R})$, so that the characteristic polynomial of L_λ is $\mu \det(L'_\lambda - \mu \operatorname{Id})$. Relations (6) and (7) become respectively

$$L'_{-\lambda} = \eta L'_\lambda \eta \tag{9}$$

and

$$(10) \qquad {}^tL'_\lambda = -\eta L'_\lambda \eta$$

where

$$\eta = \begin{pmatrix} 0 & -i & 0 & 0 \\ i & 0 & 0 & 0 \\ 0 & 0 & 0 & -i \\ 0 & 0 & i & 0 \end{pmatrix}.$$

2.2. Spectral curves and integrability

Once the trivial factor $\mu = 0$ is eliminated, the spectral curve Y has the equation $\det(L'_\lambda - \mu\,\mathrm{Id}) = 0$, that is,

$$\mu^4 - 2\left(\lambda^{-2} - 2H + 2\lambda^2\right)\mu^2 + \lambda^{-4} + 4\left(2l^2 - H\right)\lambda^{-2} + 4K = 0.$$

This is an étale double cover of the curve $C = Y/\tau_1$ whose equation is

$$\mu^4 - 2\left(z^{-1} - 2H + 2z\right)\mu^2 + z^{-2} + 4\left(2l^2 - H\right)z^{-1} + 4K = 0$$

and which is in turn a branched double cover of the curve $E = C/\tau_2$ defined by the equation

$$y^2 - 2\left(z^{-1} - 2H + 2z\right)y + z^{-2} + 4\left(2l^2 - H\right)z^{-1} + 4K = 0.$$

It goes without saying that the letters Y, C and E denote the curves once completed and normalised at $\lambda = 0$ and ∞.

Smoothness. Let us assume that $l \neq 0$. The function z is a degree-2 covering map $E \to \mathbf{P}^1(\mathbf{C})$, branched at 0 and at the roots of the ... Kowalevski polynomial φ, as the computation of the discriminant in the equation of E shows. Thus E is smooth if and only if the three roots of φ are distinct (notice that 0 is never a root of φ, since we have assumed that $l \neq 0$), that is, if and only if the point (K, H) does not belong to the cusp curve of Figure 10.

Notice that E has two points over $z = \infty$,

$$\begin{cases} (z, y) = \infty_- = (\infty, 0) \\ (z, y) = \infty_+ = (\infty, \infty) \end{cases}$$

(as usual, I find it convenient to look at the equations as if they were in $\mathbf{P}^1(\mathbf{C}) \times \mathbf{P}^1(\mathbf{C})$). The covering map $C \to E$ is branched at ∞_-, ∞_+ and also at the two points $p_1 = (\alpha, 0)$ and $p_2 = (\beta, 0)$, where I have denoted by α and β the two roots of

$$4Kz^2 + 4(2l^2 - H)z + 1 = 0.$$

Notice that α and β are mutually distinct and distinct from ∞, exactly when (K, H) is not on the parabola or the H-axis of Figure 10, so that we have eventually proved:

2.2.1 PROPOSITION. *The two curves C and Y are smooth if and only if the Kowalevski curve X is smooth.* □

In this case, C has genus 3 and Y has genus 5. Our curves (more exactly those of Bobenko, Reyman and Semenov-Tian-Shanski) are different from the genus-2 Kowalevski curve X. The relations between all these curves are still mysterious. Notice that we have concentrated on the $l \neq 0$ case. In the case of the orbit $\mathcal{O}_0$, the curve E becomes rational and C has genus 2. It seems easier to understand the relationship between C and X in this case; see the discussion in Appendix 5.

Some elements in the Picard groups. As we are going to work in $\operatorname{Pic}(C)$, it is not without point to look at some divisors on the curves E and C. First of all, it is clear that, on E,

$$\begin{aligned}(z) &= 2a - \infty_+ - \infty_- \\ (y) &= p_1 + p_2 + \infty_- - \infty_+ - 2a\end{aligned}$$

for some point a. If a_+ and a_- are the two points of C over a, this will give, on C,

$$\begin{aligned}(z) &= 2a_+ + 2a_- - 2\infty_+ - 2\infty_- \\ (\mu) &= p_1 + p_2 + \infty_- - \infty_+ - a_+ - a_-.\end{aligned}$$

Thus $2(a_+ + a_- - \infty_+ - \infty_-) = 0$ in $\operatorname{Pic}^0(C)$. Notice that y is a degree-3 meromorphic function on C, so that C cannot be a hyperelliptic curve (see e.g. Proposition III-7.10 in Farkas & Kra [27]). Thus there is no degree-2 function on C and the divisor $a_+ + a_- - \infty_+ - \infty_-$ is a non-zero element of order 2 in $\operatorname{Pic}^0(C)$.

The tangent space at any point of $\operatorname{Pic}^d(C)$ is canonically identified with the vector space $H^1(\mathcal{O}_C)$ (see Appendix 4), which we consider as the first cohomology group associated with the covering $C = \mathcal{U}_+ \cup \mathcal{U}_-$ of C (see Appendix 3). We will need the following lemma.

2.2.2 LEMMA. *The 1-cocycles defined by the holomorphic functions μ, μ^2, μ^3 on $\mathcal{U}_+ \cap \mathcal{U}_-$ have independent cohomology classes in $H^1(\mathcal{O}_C)$.*

Proof. Consider the involution $\mu \mapsto -\mu$ on C: the function μ^2 is invariant while the two others are anti-invariant,[6] so that it suffices to check that y defines a non-trivial element in $H^1(\mathcal{O}_E)$, which is obvious, and that μ and μ^2 define independent elements in $H^1(\mathcal{O}_C)$, which can be proved by a residue computation as in [15] and is left as an exercise for the reader. □

Integrability. It is an easy computation to check that we have $\operatorname{Res}[\operatorname{tr} L(\lambda)^4 \lambda^{-1} d\lambda] = 16\,[2\,(H^2+1) - K]$. Thus, using the techniques of Appendix 2 (the involution theorem A.2.1.1), we get:

2.2.3 PROPOSITION. *The two functions H and K are in involution.* □

[6] They will very soon, in this text, define elements in a Prym variety.

2.3. The eigenvector mapping

Fix an orbit $\mathcal{O}_{2l}$ ($l \neq 0$) and look at a common level $\mathcal{T}_{H,K}$ of the first integrals in $\mathcal{O}_{2l}$. These values also determine a spectral curve Y. To each element $(\Gamma, M) \in \mathcal{O}_{2l}$, associate as usual the complex line bundle F on Y, whose fibre at the point (λ, μ) is the dual of the eigenline of L_λ with respect to the eigenvalue μ.

The degree of F can be computed easily, as in Appendix 3: one uses the form (8) to see that $\lambda_\star F \to \mathbf{P}^1$ is the rank-4 trivial bundle over $\mathbf{P}^1(\mathbf{C})$, the fibre at λ being the sum of the dual vector spaces to all eigensubspaces of L_λ, that is, the whole space $\mathbf{C}^4$. One then applies the Grothendieck-Riemann-Roch theorem

$$\operatorname{ch}(\lambda_\star F)\operatorname{td}(\mathbf{P}^1) = \lambda_\star\left[\operatorname{ch}(F)\operatorname{td}(Y)\right]$$

(see A.4.1), that is

$$4(1+t) = \lambda_\star\left[(1+du)(1+(1-g)u)\right],$$

where $t \in H^2(\mathbf{P}^1)$, $u \in H^2(Y)$ are the obvious generators, d is the degree we want to compute and g, the genus of Y, is 5. Of course $\lambda_\star u = t$ and $\lambda_\star 1 = 4$ since λ has degree 4. Thus $d = 8$ and the eigenvector mapping is defined as

$$f_{H,K} : \mathcal{T}_{H,K} \longrightarrow \operatorname{Pic}^8(Y).$$

Remark. The level corresponding to the self-same values of H and K, but in $\mathcal{O}_{-2l}$, is mapped into the same Jacobian variety as above: the equation of Y depends on l only through its square l^2. However, it is clear that the two levels are isomorphic via

$$((p,q,r),(\gamma_1,\gamma_2,\gamma_3)) \longmapsto ((-p,-q,r),(\gamma_1,\gamma_2,-\gamma_3)).$$

Relation (9) implies that if v is an eigenvector of L'_λ for the eigenvalue μ then ηv is an eigenvector of L'_λ for the same eigenvalue. In particular, $\tau_1^\star F$ is isomorphic with F, and $f_{H,K}$ takes its values in the image of the mapping

$$\pi^\star : \operatorname{Pic}^4(C) \longrightarrow \operatorname{Pic}^8(Y)$$

induced by the covering map $\pi : Y \to C$.

Remark. The map $\pi^\star$ is a double cover of its image and not an inclusion as is claimed by Bobenko, Reyman and Semenov-Tian-Shanski. In terms of group homomorphisms, the kernel of its avatar

$$\pi^\star : \operatorname{Pic}^0(C) \to \operatorname{Pic}^0(Y)$$

is generated by the order-2 element $a_+ + a_- - \infty_+ - \infty_-$ that we have singled out above (see also Lemma A.5.3.2).

2.3.1 PROPOSITION. *There exists a lifting $\widetilde{f}_{H,K} : \mathcal{T}_{H,K} \to \operatorname{Pic}^4(C)$ of $f_{H,K}$ (that is, a map such that $f_{H,K} = \pi^\star \circ \widetilde{f}_{H,K}$).*

Proof. Fix a point in $\mathcal{T}_{H,K}$, in other words a matrix polynomial L'_λ. Consider a holomorphic section ψ of the associated line bundle F, or (dually) a meromorphic nowhere-vanishing section $(\lambda,\mu)\mapsto v(\lambda,\mu)$ of the eigenvector bundle. Thanks to the invariance under τ_1, and if necessary to the multiplication of v by an *ad hoc* function, one can assume that the divisor of poles of v has the form $A+\pi^{-1}(\infty_+)$ where A is an effective degree-6 divisor satisfying $\tau_1^\star A=A$, so that[7] $A=\pi^\star D$, where D is a degree-3 effective divisor on C.

The two elements of $\operatorname{Pic}^4(C)$ that are mapped to $f_{H,K}(L'(\lambda))$ are thus the classes of the divisors $D+\infty_+$ and $D+a_++a_--\infty_-$. To prove the proposition, it is enough to check that $D+a_++a_--\infty_-$ is not equivalent to an effective divisor of the form $D'+\infty_+$: it will then be possible to define $\widetilde{f}_{H,K}(L'(\lambda))$ as the class of $D+\infty_+$ (for instance).

Assume that

$$D+a_++a_--\infty_-\sim D'+\infty_+$$

or that

$$D+a_++a_-\sim D'+\infty_++\infty_-.$$

Let g be a meromorphic function on C such that $(g)=D'+\infty_++\infty_--D-a_+-a_-$. Then $\widetilde{g}=g\circ\pi$ is a function defined on Y and $\widetilde{g}\psi$ is a meromorphic section of F. Its divisor is

$$\pi^\star(D')+2\pi^{-1}\infty_++\pi^{-1}\infty_--\pi^{-1}(a_+)-\pi^{-1}(a_-).$$

As D' is effective, $\widetilde{g}\psi$ vanishes at all points over $\lambda=\infty$. This means that $\widetilde{g}\psi$ sends all the eigenvectors of L'_∞ to zero. But the latter is a diagonal matrix (in the canonical basis), so that the section $\widetilde{g}\psi$ would be zero, which is a contradiction. □

We can now state and prove the main result of this chapter.

2.3.2 Theorem. *Assume the curve C is smooth. Then the corresponding value (K,H) is regular, and $\widetilde{f}:\mathcal{T}_{H,K}\rightarrow$* $\operatorname{Pic}^4(C)$ *is an isomorphism of (complex, with real structures) algebraic varieties onto its image. This image is isomorphic to an open subset of* $\operatorname{Prym}(C|E)$.

Remark. Most of the complex aspects of this theorem are stated by Bobenko, Reyman & Semenov-Tian-Shanski [18] as a consequence of the determination of the explicit solutions of the system (E) of I.1. The theorem is obtained here independently of the solutions.

2.3.3 Corollary. *If C is smooth, the (real) level set $\mathcal{T}_{H,K}$ is diffeomorphic to the real part of* $\operatorname{Prym}(C|E)$. *In particular, the number of Liouville tori in this level is the number of connected components in the real part* $\operatorname{Prym}(C|E)_{\mathbf{R}}$.

Proof of the corollary. As $\mathcal{T}_{H,K}$ is compact, the whole real part of $\operatorname{Prym}(C|E)$ is contained in the image of $\widetilde{f}$. □

---

[7] Although the authors of [18] have been slightly less than careful on the matter in question here, this construction is taken from their paper.

Taking what we have done before into account, it is enough, in order to prove the theorem, to check that

- the tangent mapping to $f_{H,K}$ takes its values in the vector subspace of $H^1(\mathcal{O}_Y)$ consisting of elements both invariant by $\tau_1^\star$ and anti-invariant by $\tau_2^\star$: this is the tangent space to $\text{Prym}(C|E)$.
- the above tangent mapping induces an isomorphism from the tangent space to $\mathcal{T}_{H,K}$ onto this subspace.

In this way, we will know that $f_{H,K}$ (and thus $\widetilde{f}_{H,K}$ as well) is a covering map onto its image. It will then be enough to prove that $\widetilde{f}$ is injective at a generic point, or, what amounts to the same according to the factorisation through C (Proposition 2.3.1), that f has degree 2 at a generic point.

Using the techniques explained in Appendix 3, it is easy to compute the tangent mapping of the eigenvector mapping. For instance, the Hamiltonian vector field of H is mapped to the class of the cocycle defined by μ (up to a scalar constant factor) and, using the proof of Proposition 2.2.3, that of $16\,[2\,(H^2+1)-K]$ to the class of μ^3. Thus the image of the tangent mapping of f is the vector subspace of $H^1(\mathcal{O}_C)$ generated by the classes of μ and μ^3.

After Lemma 2.2.2, it is clear that the Hamiltonian vector fields of K and H are linearly independent, being mapped to independent elements of $H^1(\mathcal{O}_C)$. Thus we get:

2.3.4 PROPOSITION. *If C is smooth, the level $\mathcal{T}_{H,K}$ is regular.* □

2.3.5 PROPOSITION. *If (K,H) is such that C is smooth, the eigenvector mapping $\widetilde{f}_{H,K}$: $\mathcal{T}_{H,K} \to \text{Pic}^4(C)$ is an étale covering of its image, which is an Abelian subvariety parallel (and thus isomorphic) to* $\text{Prym}(C|E)$. □

To end the proof of Theorem 2.3.2, we still have to check that f has degree 2 (onto its image).

Once again, we use Reyman's paper [74] (see A.3.4). The inverse image of $f_{H,K}(L')$ consists of all matrices of the same form that are conjugate to L'. Now it is easy to find, for a given matrix L'_λ of type (8) associated with vectors Γ and M, all the $g \in SL(4,\mathbf{C})$ such that $gL'_\lambda g^{-1}$ is of the same type, and to deduce that

$$f^{-1}\left(f\left(\Gamma,M\right)\right)=\left\{\left(\begin{pmatrix}\gamma_1\\ \gamma_2\\ \gamma_3\end{pmatrix},\begin{pmatrix}u\\ v\\ w\end{pmatrix}\right),\left(\begin{pmatrix}\gamma_1\\ -\gamma_2\\ -\gamma_3\end{pmatrix},\begin{pmatrix}u\\ -v\\ -w\end{pmatrix}\right)\right\}$$

and thus that f has, indeed, degree 2. And this ends the proof of the theorem. □

2.4. Topology

In this subsection, I will use the real aspects of the previous results to describe the topology of the levels (K,H). From now on, everything is real.

On the one hand, from 2.3.2 we see that the critical values of (K,H) belong to the discriminant of the family of curves C (or X). On the other hand, it is not hard to find the *real* image of the momentum mapping (K,H): just consider all the components of the discriminant in the KH-plane and decide, for any of them, whether it is reached. It is obvious that the shaded regions in Figure 12 are not in the image, since they contain either points with $K < 0$ or points such that $H < -1$. To prove that the non-shaded parts of Figure 12 indeed form the image, it is sufficient to find a point of each region that is reached. This is left as an exercise for the reader.

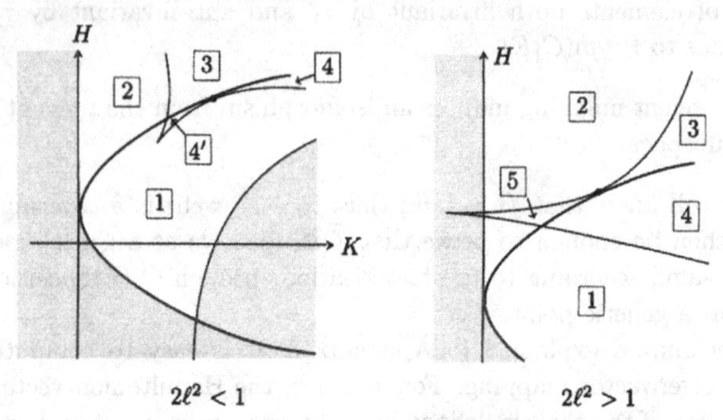

Figure 12: The image of the momentum mapping in the Kowalevski case.

All the points of the branches of curves that contribute to the boundary of the image of the momentum mapping are, of course, real critical values.

2.4.1 PROPOSITION (Kharlamov [47], Audin & Silhol [15]). *The critical values of the momentum mapping (K, H) are the points on the real part of the discriminant of the family of curves C that lie in the image of (K, H), with the exception of the branch of the parabola situated "after" its tangency with the cusp curve.*

The critical values are shown, with the image of the momentum mapping, in Figure 12. To demonstrate that the values in question are actually critical values, it is enough to find a critical point in the corresponding level. For instance, there are critical points of the form

$$(\Gamma, M) = \left(\begin{pmatrix} \gamma_1 \\ 0 \\ \gamma_3 \end{pmatrix}, \begin{pmatrix} u \\ 0 \\ w \end{pmatrix} \right)$$

mapped into all the components of the discriminant (as can be seen by direct computation), which allows us to obtain all the values claimed to be critical in Proposition 2.4.1.

It is more difficult (and rather tedious) to check directly, as did Kharlamov [47], that the points on the remaining branch of the parabola are *not* critical, but this can be obtained as a byproduct of the algebro-geometric technique used here and in [15]: the level sets here are real parts of Prym varieties that are singular[8]... but with no real singular points (the singular points come as pairs of conjugated points, of course, but they are not real).

A consequence of 2.3.3 is:

2.4.2 PROPOSITION. *The number of Liouville tori in the level corresponding to the regular value $(K, H) \in \mathbf{R}^2$ is that shown in Figure* 13.

[8]Actually compactifications of generalised Prym varieties.

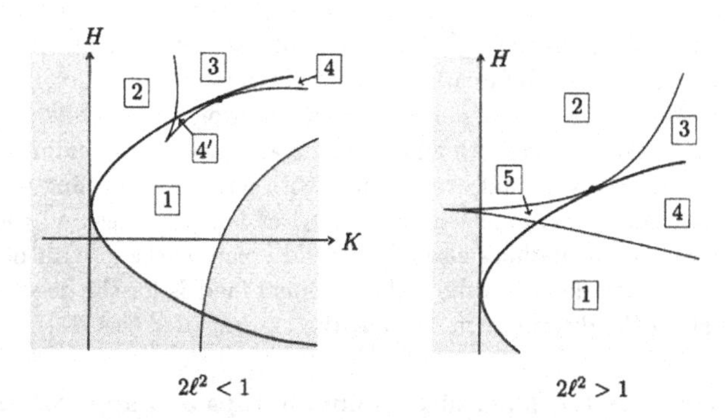

Figure 13: Liouville tori in the Kowalevski case.

This is one of Kharlamov's statements [47]. To prove it, it is enough to enumerate, in each case, the connected components of $\text{Prym}(C|E)_{\mathbf{R}}$. I send the interested reader once again to [15] for the missing details. The only trick we have used is the reduction of the enumeration of components to that of real points of order 2, using the fact that the number of real points of order 2 in a dimension-g (here $g = 2$) complex torus A is $2^g\,|\pi_0\,(A_{\mathbf{R}})|$.

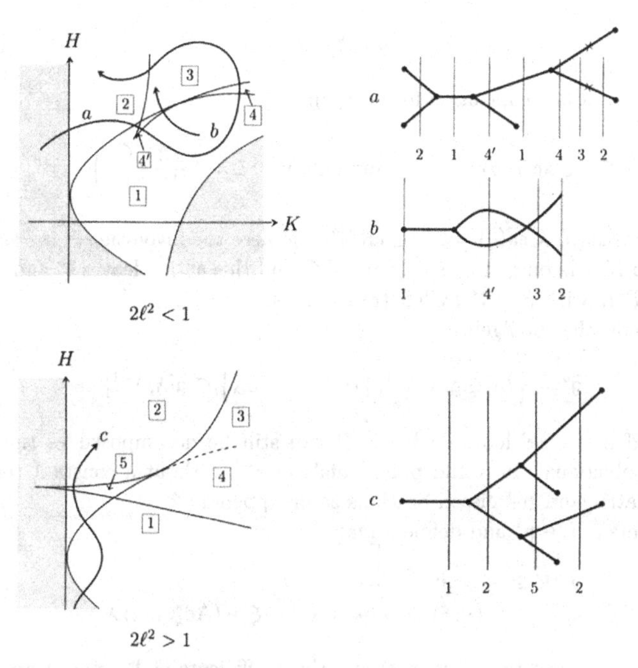

Figure 14: Bifurcations of Liouville tori in the Kowalevski case.

The same method can be used to investigate the bifurcations of the Liouville tori.

2.4.3 PROPOSITION (Kharlamov [47], Audin & Silhol [15]). *The bifurcations of the Liouville tori for the Kowalevski top are those shown on Figure* 14.

The graphs one can see in the right-hand part of the figure depict, in the manner of Fomenko (see e.g. [31]), these bifurcations (along paths labelled a, b, c).

To prove the proposition, one first notices that fibration by Liouville tori is *minimal:* the symplectic manifold $\mathcal{O}_{2l}$ is smooth and the singular fibres corresponding to the generic (regular) points of the discriminant are endowed with nowhere-vanishing vector fields, so that they must be unions of immersed circles, tori or Klein bottles and thus cannot be obtained by blowing up something else (this would create surfaces with non-zero Euler characteristic). One then describes the stable minimal models for the degeneracies of the Prym varieties along the discriminant. This is the hardest part (see [15]).

3. Lax pairs for generalised spinning tops and applications

3.1. Lax pairs for spinning tops

In this subsection, I will explain briefly how the Lax pair I have just used was found. I follow closely the original paper [18] of Bobenko, Reyman and Semenov-Tian-Shanski. The situation[9] is that of a Lie group G (think of $G = SO(p,q)$) with Lie algebra $\mathfrak{g}$, endowed with an involution σ whose subgroup of fixed points will be denoted by K (think of $K = SO(p) \times SO(q)$). At the Lie algebra level, the fixed and anti-fixed points of σ (I use the same letter σ for the involutions on G and on $\mathfrak{g}$) give a decomposition

$$\mathfrak{g} = \mathfrak{k} + \mathfrak{p}$$

(think of the block decomposition for $\mathfrak{so}(p,q)$:

$$L \in \mathfrak{so}(p,q) = \mathfrak{g} \quad \text{if and only if} \quad L = \left(\begin{array}{c|c} A & C \\ \hline {}^tC & B \end{array}\right)$$

where $(A, B) \in \mathfrak{so}(p) \times \mathfrak{so}(q) = \mathfrak{k}$, $c \in (\mathbf{R}^p)^q = \mathfrak{p}$; here the involution σ is $\sigma(L) = -{}^tL$).

The group K acts on $\mathfrak{g}^\star$ as a subgroup of G and this action leaves $\mathfrak{k}^\star$ and $\mathfrak{p}^\star$ invariant. We identify $T^\star K$ with $K \times \mathfrak{k}^\star$ by left translations.

Consider now the subalgebra

$$\tilde{\mathfrak{g}}^\sigma = \left\{L_\lambda \in \mathfrak{g}[\lambda, \lambda^{-1}] \mid \sigma(L_\lambda) = L_{-\lambda}\right\} \subset \mathfrak{g}[\lambda, \lambda^{-1}]$$

(this is called a *twisted* loop algebra). It can still be decomposed as the sum of the algebras of polynomials in λ and polynomials in λ^{-1} without a constant term, thus we have an R-matrix and R-Poisson brackets as in Appendix 2.

Fix elements a, $h \in \mathfrak{p}^\star$ and define a map

$$\begin{array}{rcl} T^\star K \cong K \times \mathfrak{g}^\star & \longrightarrow & \tilde{\mathfrak{g}}^\sigma \\ (k, \xi) & \longmapsto & h\lambda^{-1} + \xi + (\operatorname{Ad}^\star_{k^{-1}} \cdot a)\,\lambda \end{array}$$

(note that the constant term ξ is in $\mathfrak{k}^\star$ and the coefficients of λ^{-1} and λ are in $\mathfrak{p}^\star$). One then checks that

3.1.1 LEMMA. *The map $T^\star K \to \tilde{\mathfrak{g}}^\sigma$ is a Poisson map.*

[9] I will explain the construction in full generality and illustrate it at each step by an example relevant for spinning tops.

See the original paper [18] of our three authors if necessary. Using the R-matrix/AKS technique described in Appendix 2 we know that we have constructed Hamiltonian systems with many first integrals.

It is convenient to make, as in the original paper, a small (symplectic) change of variables $(k,\xi) \mapsto (k^{-1}, -\operatorname{Ad}_k^\star \cdot \xi)$ so that the new "matrix" has the form

$$h\lambda^{-1} - \operatorname{Ad}_k^\star \cdot \xi + (\operatorname{Ad}_k^\star \cdot a)\,\lambda.$$

In the concrete example $G = SO(p,q)$, one gets eventually:

3.1.2 THEOREM (Bobenko, Reyman & Semenov-Tian-Shanski [18]). *Let A and H be two fixed elements in $(\mathbf{R}^p)^q$ ($p \times q$ matrices). The coefficients in the characteristic polynomial of the matrix*

$$L_\lambda = \left(\begin{array}{c|c} 0 & H \\ \hline {}^tH & 0 \end{array}\right)\lambda^{-1} - \left(\begin{array}{c|c} k\xi^t k & 0 \\ \hline 0 & r\omega^t r \end{array}\right) + \left(\begin{array}{c|c} 0 & kA^t r \\ \hline r^tA^t k & 0 \end{array}\right)\lambda$$

are functions of

$$(k,r,\xi,\omega) \in SO(p) \times SO(q) \times \mathfrak{so}(p)^\star \times \mathfrak{so}(q)^\star \cong T^\star\left(SO(p) \times SO(q)\right);$$

which commute with respect to the canonical Poisson structure. □

Note that the matrix L_λ is invariant by the right action of the stabiliser K_A of A in K. Suppose $q \leq p$ and let A be the matrix consisting of the first q vectors of the canonical basis of $\mathbf{R}^p$. Then K_A consists of the matrices

$$\left(\begin{array}{c|c} k & 0 \\ \hline 0 & r \end{array}\right)$$

such that $kA^t r = A$, that is, in this case, such that

$$k = \left(\begin{array}{c|c} r & 0 \\ \hline 0 & s \end{array}\right)$$

for an arbitrary $s \in SO(p-q)$. Hence K_A is isomorphic to $SO(q) \times SO(p-q)$, and is embedded as a subgroup in $SO(p) \times SO(q)$ by the map

$$(r,s) \longmapsto \left(\begin{array}{c|c|c} r & 0 & 0 \\ \hline 0 & s & 0 \\ \hline 0 & 0 & r \end{array}\right).$$

Let us perform the reduction with respect to the action of the diagonal subgroup $SO(q)$ ($s = 1$). The momentum mapping $\mu : T^\star K \to \mathfrak{so}(q)^\star$ for this action is

$$(k,r,\xi,\omega) \longmapsto \xi' + \omega$$

where ξ' is the obvious truncation of the $p \times p$ skew-symmetric matrix ξ.

Our reduced phase space is isomorphic to $T^\star SO(p)$. Note that, even if the former does not depend on q any more, the first integrals we obtain from Theorem 3.1.2 do depend on q. The reduction of the Hamiltonian

$$H = \operatorname{Res}\operatorname{tr}\left(L_\lambda^2 \lambda^{-1} d\lambda\right)$$

is

$$H = \sum_{i,j=1}^{p} \xi_{i,j}^2 + \sum_{i,j=1}^{q} \xi_{i,j}^2 - \sum_{i=1}^{q} (ke_i \cdot H_i)$$

with more or less obvious notation for the potential term here. The interesting part is actually the quadratic part, the kinetic energy. Note that the doubling of the terms $\xi_{i,j}^2$ $(i, j \leq q)$ comes from the reduction process, so that this process is responsible for our obtaining a Kowalevski-type Hamiltonian (remember the form of the inertia matrix in I.2.1).

The case of the Kowalevski top corresponds to $(p, q) = (3, 2)$. The $(p, q) = (3, 1)$ case gives a Lax matrix for the Lagrange top (see 3.2).

Returning to more general Lie groups, the case where $G = SL(n, \mathbf{R})$ and $K = SO(n)$ gives the (Manakov) free n-dimensional rigid body (see Chapter IV). Reyman and Semenov-Tian-Shanski were even able to use $G = G_2$ to get an exotic system on $SO(4)$ (see [76]).

3.2. The symmetric top revisited

Let us come back to the case of the symmetric spinning top. It is described by $p = 3$, $q = 1$ in the construction above. Using the notation of section 2 together with that of Chapter I:

3.2.1 PROPOSITION (Bobenko, Reyman & Semenov-Tian-Shanski [18]). *In the Lagrange case, system (E) is equivalent to the Lax equation*

$$\frac{d}{dt} L_\lambda = [L_\lambda, M_\lambda]$$

where

$$L_\lambda = \begin{pmatrix} 0 & \Gamma \\ {}^t\Gamma & 0 \end{pmatrix} \lambda^{-1} + \begin{pmatrix} -M & 0 \\ 0 & PMP \end{pmatrix} + \begin{pmatrix} 0 & L \\ {}^tL & 0 \end{pmatrix} \lambda$$

and M_λ is the polynomial part $(L_\lambda^2)_+$ of L_λ^2.

Remark. In the block decomposition of $\mathfrak{so}(3,1)$, any matrix is written as

$$M = \begin{pmatrix} A & v \\ {}^tv & 0 \end{pmatrix}$$

where $A \in \mathfrak{so}(3)$ is a skew-symmetric matrix and $v \in \mathbf{R}^3$ is a vector, so that, as a vector space, $\mathfrak{so}(3,1) = \mathfrak{so}(3) \times \mathbf{R}^3$. The Lie algebra structure is given by the bracket

$$[(A, v), (B, w)] = ([A, B] + v \wedge w, A \cdot w - B \cdot v).$$

Note that this Lie bracket is different from the Lie bracket associated with the Lie algebra of the group of rigid motions used in I.1.2.

The spectral curve(s). The spectral curve Y has the equation

$$\mu^4 + \left[H' - \left(\lambda^{-2} + \lambda^2\right)\right]\mu^2 - \left(c\lambda^{-1} + K\lambda\right)^2 = 0$$

(normalised as usual at $\lambda = 0$ and $\lambda = \infty$). It has several involutions, among them $(\mu, \lambda) \mapsto (\mu, -\lambda)$, that gives as quotient the curve C:

$$\mu^4 + \left[H' - \left(z^{-1} + z\right)\right]\mu^2 - z^{-1}\left(c + Kz\right)^2 = 0.$$

Consider the degree-2 map $\mu : C \to \mathbf{P}^1$. Its branching points are the roots of the discriminant of the equation for C, viewed as a degree-2 equation in z. Here the discriminant has the form

$$\Delta(\mu) = \mu^2\delta(\mu)$$

so that 0 is always (i.e. for all values of H and K) a double root of Δ. This means that C always has a double point, so that the spectral curve Y always has two double points.

The main reason why I wanted to discuss this example is that it shows that there is no hope of proving a general statement of equivalence between the smoothness of the spectral curve and the regularity of the levels.

Remark. However, one can separate the two branches at a double point of C to get generically smooth curves $\widetilde{Y}$ and $\widetilde{C}$. The latter has genus 2 and is a double covering of E branched at two points. Looking carefully at the construction of the eigenvector mapping, in order to generalise it slightly, it is then possible to map the corresponding level into $\operatorname{Pic}(\widetilde{Y})$. The same symmetry considerations used in 2.3 show that this generalised eigenvector mapping can be modified to define a map into $\operatorname{Pic}(\widetilde{C})$, taking its values in $\operatorname{Prym}(\widetilde{C}|E)$. This is now a dimension-1 Abelian variety, which must be related to the genus-1 curve X we had in Chapter II.

3.3. A curiosity, the Goryachev-Chaplygin case

The R-matrix/AKS-approach (Appendix 2) is very powerful for *constructing* integrable systems. However, it can be very hard to *put a given system into Lax form*. If it is understandable that the methods of Bobenko, Reyman & Semenov-Tian-Shanski [18] explained above give Lax pairs for the Lagrange and Kowalevski tops, it is more surprising that they give a mysterious byproduct: Bobenko and Kuznetsov derive in [19] a Lax pair for the Goryachev-Chaplygin case. This really looks like a cookery recipe: take an avatar of the Lax pair that works for Kowalevski, remove the first column and the first row of both matrices ... and that is all.

A good avatar to use is the one in equation (8), slightly transformed by an *ad hoc* conjugation (the process is, of course, very sensitive to conjugation), which gives

$$L_\lambda = \begin{pmatrix} 0 & -i\gamma_3\lambda^{-1} & -v + iu \\ i\gamma_3\lambda^{-1} & -2iw & (\gamma_2 - i\gamma_1)\lambda^{-1} - 2i\lambda \\ v + iu & (\gamma_2 + i\gamma_1)\lambda^{-1} + 2i\lambda & 2iw \end{pmatrix}$$

and

$$M_\lambda = \begin{pmatrix} 3iw & 0 & v - iu \\ 0 & 2iw & 2i\lambda \\ -v - iu & -2i\lambda & -2iw \end{pmatrix}.$$

3.3.1 PROPOSITION (Bobenko & Kuznetsov [19]). *In the Goryachev-Chaplygin case, system (E) of I.1 is equivalent to*

$$\frac{d}{dt}L_\lambda = [L_\lambda, M_\lambda].$$

The proof of this proposition is, once again, straightforward and left as an exercise for the reader. From this Lax pair, one gets a spectral curve Y:

$$\det(L_\lambda - \mu \operatorname{Id}) = -\mu^3 + (\lambda^{-2} - 2H + 4\lambda^2)\mu - 2iK$$

endowed with an involution ($\lambda \mapsto -\lambda$), with quotient X:

$$-\mu^3 + (z^{-1} - 2H + 4z)\mu - 2iK = 0.$$

In [19], the solutions of system (E) are derived. It turns out that they can be expressed using *square roots* of ϑ-functions associated with the Jacobian of X (note that this means that the flows on the Jacobian are not linear). Actually, it is well known that, in this case, the common levels of H and $K$are *branched* double covers of Abelian surfaces (see e.g. Piovan's paper [71]). Note that this is coherent with the following general principle: if the Lax pair you are using was obtained in an honest way, the flow it describes is linearised on an Abelian variety and your level has a compactification that is an (étale) covering of this Abelian variety. (I am indebted to Alexei Reyman for this remark, which is the philosophical foundation of his results, quoted in Appendix 3.)

From the topological point of view, the Lax equation is not very good. If the curve X carries a real structure (change μ to $i\mu$ to make it more apparent) and is equivalent to the curve used by Goryachev and Chaplygin, it might be true but is not obvious that the matrix L would give a *real* eigenvector mapping.

IV

The free rigid body

We have already discussed the example of the free rigid body in Chapter I: this gives Euler's equations $\dot{M} = [M, \Omega]$. In this chapter, I want to show how the method using eigenvectors of Lax matrices can be used to prove the results already obtained (the algebraic content is even more visible here). I will then describe an analogous question, that of the so-called dimension-4 free rigid body, which is an integrable system on *compact* four-dimensional symplectic manifolds, the coadjoint orbits of $SO(4)$. The algebraic aspects are very rich and I will not try to discuss them all.

1. The Euler and Manakov equations

1.1. The Euler equations

Consider system (E) of I.1 in the absence of gravity, namely $\dot{M} = [M, \Omega]$ where M and Ω are two 3×3 skew-symmetric matrices satisfying, as vectors of $\mathbf{R}^3$, the relation $M = \mathfrak{J}\Omega$.

One is of course tempted to replace $\mathfrak{so}(3)$ by $\mathfrak{so}(n)$, which is possible provided that the relation $M = \mathfrak{J}\Omega$ is rewritten in terms of 3×3 matrices, in order to generalise it to $n \times n$ matrices. This is easily done by the introduction of a diagonal matrix J such that $M = \mathfrak{J}\Omega$ is equivalent to $M = \Omega J + J\Omega$: if $\mathfrak{J}$ is the diagonal matrix with entries a_i ($\lambda_i = 1/a_i$), then J is the diagonal matrix of the b_i where

$$\begin{cases} a_1 &= b_3 + b_2 \\ a_2 &= b_1 + b_3 \\ a_3 &= b_2 + b_1. \end{cases}$$

The general system in $\mathfrak{so}(n)$,

$$\begin{cases} \dot{M} &= [M, \Omega] \\ M &= \Omega J + J\Omega, \end{cases}$$

is called the Euler-Arnold equations.

1.2. The Manakov equations

Notice that $M = [M, \Omega]$ is a Lax equation. However, we have seen in Chapter I that the solutions in the $n = 3$ case are elliptic functions, so that a Lax form without spectral parameter cannot be used to solve the equations.

Since $[\Omega J + J\Omega, J] = \Omega J^2 - J^2\Omega$, it follows that $[M + J^2\lambda, \Omega + J\lambda] = [M, \Omega]$ so that the Euler-Arnold equations are equivalent to Manakov's equation [60]:

$$\frac{d}{dt}\left(M + J^2\lambda\right) = [M + J^2\lambda, \Omega + J\lambda], \tag{1}$$

a Lax equation with a spectral parameter.

Though found thirteen years earlier, this Lax equation belongs to the family that was described[1] in [18] by Bobenko, Reyman and Semenov-Tian-Shanski and in the present III.3. It seems to have played a precursory role in the theory.

2. The dimension-3 free rigid body

We use Manakov's equation to re-prove the results of I.3. With the same notation for J and $\mathfrak{J}$ as above, notice that the conditions $0 < \lambda_1 < \lambda_2 < \lambda_3$ or $0 < a_3 < a_2 < a_1$ of Chapter I give $b_1 < b_2 < b_3$. Assume moreover as in I.1 that $a_1 < a_2 + a_3$, so that $b_1 > 0$ and $0 < b_1^2 < b_2^2 < b_3^2$.

Remark. To my knowledge, this is the only place in this book where the above condition, imposed by physics on the matrix $\mathfrak{J}$ is used albeit weakly, in a mathematical treatment of the problem.

2.1. The spectral curve

Let us compute the characteristic polynomial of $M + J^2\lambda$. As J^2 is a diagonal matrix, let us instead call the eigenvalue $\lambda\mu$, so that $J^2 - \mu\,\mathrm{Id}$ is the diagonal matrix whose entries are $\beta_i = b_i^2 - \mu$ and the characteristic polynomial is

$$\det\left[M + \lambda\left(J^2 - \mu\,\mathrm{Id}\right)\right] = \lambda\left(\beta_1\beta_2\beta_3\lambda^2 + \beta_1 u^2 + \beta_2 v^2 + \beta_3 w^2\right).$$

We shall consider the spectral curve X defined by the equation

$$\lambda^2 + \frac{u^2}{(b_2^2 - \mu)(b_3^2 - \mu)} + \frac{v^2}{(b_3^2 - \mu)(b_1^2 - \mu)} + \frac{w^2}{(b_1^2 - \mu)(b_2^2 - \mu)} = 0,$$

which can also be written

$$\lambda^2 + \frac{2(H' - p^2\mu)}{(b_1^2 - \mu)(b_2^2 - \mu)(b_3^2 - \mu)} = 0 \tag{2}$$

so that $2p^2 = u^2 + v^2 + w^2$ (using the same notation as in I.3) and $2H' = b_1^2 u^2 + b_2^2 v^2 + b_3^2 w^2$. The function H' can of course be written as a linear combination of H and $K = p^2$, but this is left as an exercise for the reader.

The genus of the curve X is 1 and the curve is a double cover of $\mathbf{P}^1$ (this is what the function μ does) branched at the points b_i^2 and H'/p^2. Set $P_i = (b_i^2, \infty)$ and notice that all the branching points are real and, thus, that the real part $X_{\mathbf{R}}$ has two connected components.

---

[1] Here $G = SL(n, \mathbf{R})$, $K = SO(n)$ and σ is the Cartan involution.

2.2. The eigenvector mapping

I will continue to use the convenient notation $\beta_i = b_i^2 - \mu$. The vector

$$W(\mu,\lambda) = \begin{pmatrix} \dfrac{-\lambda\beta_2 v + uw}{\lambda^2\beta_1\beta_2 + w^2} \\[2ex] \dfrac{\lambda\beta_1 u + vw}{\lambda^2\beta_1\beta_2 + w^2} \\[2ex] 1 \end{pmatrix}$$

is an eigenvector of our Lax matrix $M + J^2\lambda$ (by direct computation), a meromorphic section of the eigenvector bundle over X (see Appendix 3). It is chosen to have no zero on X, but it has, of course, poles. Its pole divisor represents an element of the Picard group $\operatorname{Pic}(X)$ that correspond to the (dual of the) eigenvector bundle.

2.2.1 PROPOSITION. *The pole divisor of W is $P_1 + P_2 + A \in \operatorname{Div}^3(X)$, where A is the point of X having coordinates*

$$\left(\mu = \frac{b_1^2 u^2 + b_2^2 v^2}{u^2 + v^2}, \quad \lambda = -\frac{uw}{(b_2^2 - y)\,v}\right).$$

Proof. Let us look for the poles of the first coordinate of W, that is, for the meromorphic function$(-\lambda\beta_2 v + uw) / (\lambda^2\beta_1\beta_2 + w^2)$.

We consider first the points at infinity P_1, P_2 and P_3 on X. At P_1, the numerator is infinite, but not the denominator, so that P_1 is a pole. At P_2, the numerator vanishes, at P_3 it goes to infinity but the quotient is finite. Let us look now for the "finite" poles, that is, the points at which the denominator vanishes but not the numerator. Now, $\lambda^2\beta_1\beta_2 + w^2 = 0$ if and only if $u^2\beta_1 + v^2\beta_2 = 0$, which gives us one value of μ, that in the statement of the proposition, corresponding to two points of X, that is, two opposite values of λ, at one of which the numerator will vanish as well:

$$0 = \lambda^2\beta_1\beta_2 + w^2 = -\lambda^2\beta_2^2\frac{v^2}{u^2} + w^2 = \left(-\lambda\beta_2\frac{v}{u} + w\right)\left(\lambda\beta_2\frac{v}{u} + w\right)$$

since $\beta_1 u^2 + \beta_2 v^2 = 0$. We are left with the point A.

The same computation shows that this is also a pole of the second coordinate

$$\frac{\beta_1 u + vw}{\lambda^2\beta_1\beta_2 + w^2},$$

as is P_2. In addition, it shows that all these poles are simple. Hence the result. □

I cannot refrain from translating by the constant divisor $-P_1 - P_2$, obtaining the following map to $\operatorname{Pic}^1(X) = X$:

$$\begin{array}{ccc} V & \xrightarrow{\varphi} & X \\ M & \longmapsto & A(M). \end{array}$$

Recall from I.3 that V is the name of a completion of the common level of H and K that we have proved to be a genus-1 curve by virtue of being the intersection of two quadrics. The next proposition, in a slightly different language, is due to Haine [39].

2.2.2 PROPOSITION. *The genus-1 curves V and X are isomorphic, but the map φ is an étale covering map of group $\mathbf{Z}/2 \times \mathbf{Z}/2$.*

Remark. This tells us among other things that X and V are smooth at the same time.

Proof. The mapping φ is well defined: the numerator and the denominator cannot both vanish[2] if (u, v, w) satisfy the equations of V. We thus have an algebraic map from the affine part $V_0 \subset \mathbf{C}^3 \subset \mathbf{P}^3(\mathbf{C})$ to the curve defined by equation (2) in $\mathbf{P}^1 \times \mathbf{P}^1$. This defines an algebraic mapping $V \longrightarrow X$.

Clearly one can reconstruct u^2, v^2 and w^2 from y using the equations for V. Knowledge of λ gives the sign of uw/v, so that φ has degree 4. The inverse image of a point of X consists of the points

$$\begin{pmatrix} \varepsilon u \\ \eta v \\ \varepsilon\eta w \end{pmatrix} \qquad \varepsilon, \eta = \pm 1.$$

One can say more: these four points are always distinct, otherwise two of the coordinates u, v, w would vanish at the same time, which we have already noticed to be impossible. Hence we have an étale degree-4 covering map with group $\mathbf{Z}/2 \times \mathbf{Z}/2$.

Thanks to this last point, the curves V and X will nevertheless be isomorphic: both V and X are actually groups (Abelian varieties) and the algebraic map φ must be a group homomorphism (up to a translation) with kernel isomorphic to $\mathbf{Z}/2 \times \mathbf{Z}/2$, from which it is deduced that V is indeed isomorphic to X and φ is (up to a translation) multiplication by two. Here are the details. First, a rather general rigidity lemma coming from Mumford's book on Abelian varieties [64].

2.2.3 LEMMA. *Let V and X be two Abelian varieties, and let $\varphi : V \longrightarrow X$ be a morphism of varieties. There exists $a \in X$ such that $\varphi = f + a$ where $f : V \longrightarrow X$ is a morphism of groups.*

Proof. Let us put $f(x) = \varphi(x) - \varphi(0)$ so that $f(0) = 0$. We want to show that f is a morphism of groups, in other words that

$$\begin{aligned} g : V \times V &\longrightarrow X \\ (x, y) &\longmapsto f(x+y) - f(x) - f(y) \end{aligned}$$

is constant (actually zero). Notice that $g(x, 0) = g(0, y) = 0 \ \forall x, y \in V$. Let U be an affine chart in a neighbourhood of $0 \in X$ and let $F = X - U$ be its complement. The set $g^{-1}(F)$ is closed, as is its projection G on the second factor V: the projection map is closed because V is compact. By definition, 0 does not belong to G so that the complement $V - G$ is a non-empty open subset Ω of V. Let $y \in \Omega$. Then $g\,|_{V \times \{y\}}$ is a morphism from the compact variety V into the open affine set U. It must be constant and equal to zero. We thus have shown that g is constant on the non-empty open subset $V \times \Omega \subset V \times V$, and thus that it is constant. □

Let us come back to the curves V and X. Identify $V = \mathbf{C}/\Lambda_1$ and $X = \mathbf{C}/\Lambda_2$ in such a way that $\varphi(0) = 0$ (in other words the translation is incorporated in the identification). Hence φ is a group homomorphism that can be lifted as $\tilde{\varphi} : \mathbf{C} \longrightarrow \mathbf{C}$, of the form $z \longmapsto Az$ for some $A \in \mathbf{C}$. As $\operatorname{Ker} \varphi = \mathbf{Z}/2 \times \mathbf{Z}/2$ one can assume that the two lattices coincide and that $A = 2$ (details are given in the proof following 2.2.4). □

[2]If $b_1 \neq b_2$, as we have assumed.

Remark. It would of course be possible to prove that V and X are isomorphic using the relations between the λ_i and the b_i^2 but this would not be much fun (to tell the truth, I find the idea so unappealing that I have never even tried it).

If we are more careful with the real structures in the last part of the proof, we will get:

2.2.4 PROPOSITION. *The curve X is isomorphic to the Jacobian of V. The isomorphism is real. The eigenvector mapping φ is a covering map of group $\mathbf{Z}/2 \times \mathbf{Z}/2$. When $V_{\mathbf{R}}$ is non-empty, it is mapped onto one connected component of $X_{\mathbf{R}}$ and defines a double covering of each component of $V_{\mathbf{R}}$ onto its image.*

Proof. Notice that $X_{\mathbf{R}}$ is always non-empty, but that this is not the case for $V_{\mathbf{R}}$. If $V_{\mathbf{R}}$ is empty, that is, if $h/p^2 \notin [\lambda_1, \lambda_3]$, just replace V by its Jacobian and extend the map φ. We thus consider only the case where $V_{\mathbf{R}}$ is non-empty so that both the identifications $V = \mathbf{C}/\Lambda_1$ and $X = \mathbf{C}/\Lambda_2$ above can be made real. As the two curves each have a real part with two components, it may be assumed that the lattice Λ_j is generated by 1 and a pure-imaginary τ_j (see Appendix 4). As φ is a real map, $A = \widetilde{\varphi}(1) \in \left(\mathbf{R} \cup \mathbf{R} \cdot \frac{1}{2}\tau_2\right) \cap \Lambda_2$, that is $A \in \mathbf{Z}$. Of course, if $A = n$, there is an order-n element in $\operatorname{Ker}\varphi$ so that $A = 2$. Now $\widetilde{\varphi}(\tau_1) = 2\tau_1 = q\tau_2$ for some $q \in \mathbf{Z}$. To get the required kernel we must have $q = 2$ and $\tau_1 = \tau_2$.

Now we know everything quite explicitly, for instance that the restriction of φ to the real part of V (if non-empty) sends the two components of $V_{\mathbf{R}}$ onto the same component of $X_{\mathbf{R}}$ and that its restriction to each component of $V_{\mathbf{R}}$ is a double covering map. □

Remark. The component of $X_{\mathbf{R}}$ in the image is the one that contains the point $(\mu = H'/p^2, \lambda = 0)$, obtained for $w = 0$.

3. Remarks on the dimension-4 rigid body

Consider now Manakov's equation (1) in $\mathfrak{so}(4)$. The entries of the diagonal matrix J (see section 1) will be denoted by b_i ($0 \le i \le 3$); we will assume that they are distinct and that they satisfy $b_0^2 < b_1^2 < b_2^2 < b_3^2$.

3.1. The symplectic orbits and the first integrals

Let us write the skew-symmetric matrix M as

$$M = (x, y) = \begin{pmatrix} 0 & -x_3 & x_2 & y_1 \\ x_3 & 0 & -x_1 & y_2 \\ -x_2 & x_1 & 0 & y_3 \\ -y_1 & -y_2 & -y_3 & 0 \end{pmatrix}$$

and compute the characteristic polynomial, $\det(M + \lambda J^2 - \lambda\mu \operatorname{Id})$ (as above, the eigenvalue is $\lambda\mu$), that is,

$$\lambda^4 \prod_{i=0}^{3} \left(b_i^2 - \mu\right) + \lambda^2 \left[\mu^2 f_1(x,y) - \mu H(x,y) + K(x,y)\right] + f_2(x,y)^2$$

where

$$\begin{aligned}
f_1(x,y) &= \sum x_i^2 + \sum y_i^2 \quad \text{is the } \textit{norm} \text{ of } M,\\
f_2(x,y) &= \sum x_i y_i \quad \text{is its } \textit{Pfaffian},\\
H(x,y) &= \sum_{i=1}^{3} \left(b_i^2 + b_0^2\right) x_i^2 + \frac{1}{2} \sum_{\{i,j,k\}=\{1,2,3\}} \left(b_i^2 + b_j^2\right) y_k^2,\\
K(x,y) &= \sum_{i=1}^{3} \left(b_i^2 b_0^2\right) x_i^2 + \frac{1}{2} \sum_{\{i,j,k\}=\{1,2,3\}} \left(b_i^2 b_j^2\right) y_k^2.
\end{aligned}$$

Recall that the Pfaffian of a skew-symmetric matrix is the evaluation on its entries of a universal polynomial whose square is the determinant (see e.g. Bourbaki's algebra textbook [20]). For instance, $f_2 = 0$ if and only if the rank of M is ≤ 2.

It is obvious that f_1 and f_2 are invariant under conjugation by orthogonal matrices, and thus are orbit invariants for the adjoint or coadjoint actions of $SO(4)$. It is classical (and left to the reader as an exercise) that they are actually *the* orbit invariants, and thus the Casimirs for the canonical Poisson structure (see Appendix 1).

Remark. It is indeed the Pfaffian f_2 that is an orbit invariant, although its square, the determinant, is what appears in the characteristic polynomial. The situation is very similar to that in III.2.3. We fix an orbit $\mathcal{O}_f$, on which we look at a common level of H and K: $\mathcal{T}_h$ will thus be a common level of f_1, f_2, H and K, even if the curve C determines *only* f_2^2.

If we now define $u_i = x_i + y_i$ and $v_i = x_i - y_i$, we get

$$\sum u_i^2 = f_1 + 2f_2 \quad \text{and} \quad \sum v_i^2 = f_1 - 2f_2.$$

Thus, if $f_1 > |2f_2|$, the orbit $\mathcal{O}_f$ indexed by f_1 and f_2 is diffeomorphic to $S^2 \times S^2$. Thus we have defined the generic orbits. If $f_1 = |2f_2|$, the orbit is a 2-sphere.

We will consider only the generic orbits $\mathcal{O}_f$ (with $f = (f_1, f_2)$ and $f_1 > |2f_2|$); in particular, from now on, $f_1 \neq 0$. The two functions H and K are then first integrals.

3.1.1 Proposition. *The functions H and K are in involution.*

Proof. For any $n \times n$ matrix A, let $(-1)^k \sigma_k$ be the coefficient of y^{n-k} in $\det(y \operatorname{Id} - A)$. This defines an invariant function σ_k on the Lie algebra $\mathfrak{gl}(n, \mathbf{C})$ of all matrices. Our first integrals are related to this construction, since $H = \operatorname{Res}\left[\sigma_4 \left(M + \lambda^2 J\right) \lambda^{-3} d\lambda\right]$ and $K = \operatorname{Res}\left[\sigma_3 \left(M + \lambda^2 J\right) \lambda^{-2} d\lambda\right]$ (by direct computation), so that they commute according to the involution theorems described in Appendix 2. □

The system is thus integrable. After having performed a translation $\mu \mapsto \mu - b_0^2$ if necessary, we may assume that $b_0^2 = 0$. The first integrals then have the very simple form

$$\begin{aligned}
H(x,y) &= \sum_{i=1}^{3} x_i^2 + \frac{1}{2} \sum_{\{i,j,k\}=\{1,2,3\}} \left(b_i^2 + b_j^2\right) y_k^2,\\
K(x,y) &= \frac{1}{2} \sum_{\{i,j,k\}=\{1,2,3\}} b_i^2 b_j^2 y_k^2.
\end{aligned}$$

Remark. The four functions f_1, f_2, H, K are *quadratic* in the entries of M, so that a common level set is an intersection of quadrics. Such algebraic varieties carry a very rich structure, similarly to the lower-dimensional example of the curve V used in I.3.2 and in section 2. A lot of work has been done on this aspect, e.g. by Adler & van Moerbeke [3]. An "exotic" Hamiltonian system on $\mathfrak{so}(4)^\star$ with a quadratic Hamiltonian and a degree 4 second first integral was constructed by Reyman and Semenov-Tian-Shanski (see [76]) as a byproduct of their constructions with Bobenko in [18] (see III.3).

3.2. The spectral curves

The spectral curve is a curve C having equation

$$C: \qquad \lambda^4 \prod_{i=0}^{3}\left(b_i^2-\mu\right)+\lambda^2\left(\mu^2 f_1-\mu H+K\right)+f_2^2=0$$

a double cover $p: C \to E$ of a curve E obtained on dividing by the involution $\lambda \mapsto -\lambda$,

$$E: \qquad z^2 \prod_{i=0}^{3}\left(b_i^2-\mu\right)+z\left(\mu^2 f_1-\mu H+K\right)+f_2^2=0$$

$(z=\lambda^2)$.

Assume first that $f_2 \neq 0$. The genus of the curve E is then 1, as E is a double cover of $\mathbf{P}^1$ branched at the four roots of

$$\Delta(\mu)=\left(\mu^2 f_1-\mu H+K\right)^2-4 f_2^2 \prod_{i=0}^{i=4}\left(b_i^2-\mu\right).$$

Let us write, to simplify the notation,

$$f_1^2-4 f_2^2=f_1^2\left(1-\alpha^2\right) \quad \text{with} \quad \alpha=\frac{|2 f_2|}{f_1} \in]0,1[$$

and $h=H/f_1$, $k=K/f_1$. A more significant notation is

$$w^2=\prod_{i=0}^{i=4}\left(b_i^2-\mu\right)$$

so that (μ, w) is a point on a genus-1 curve $\mathcal{E}$ (Figure 15). Notice that we have two distinct elliptic curves playing different roles here.

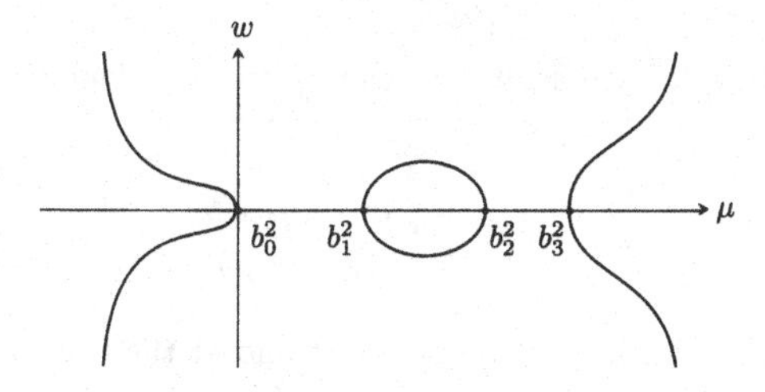

Figure 15: The real part of $\mathcal{E}$.

Now, $\Delta(\mu)=f_1^2 \delta(\mu)$ where

$$\delta(\mu)=\left(\mu^2-\mu h+k\right)^2-\alpha^2 w(\mu)^2.$$

The covering $C \to E$ is branched at the four points P_i: $(z, \mu) = (\infty, b_i^2)$ of E. Since we have assumed these points to be distinct, it has genus 3 and is smooth if and only if E is smooth, that is, if the four roots of δ are distinct. Now μ is a double root of δ if and only if $\delta(\mu) = \delta'(\mu) = 0$, that is, if and only if

$$\begin{cases} h = 2\mu + \alpha w'(\mu) \\ k = \mu^2 - \alpha\left[w(\mu) - \mu w'(\mu)\right] \\ \text{with} \quad w^2 = \prod (b_i^2 - \mu). \end{cases}$$

These equations are a parametrisation of the discriminant of the family of curves E obtained as h and k vary.

Remark. Even if one is interested only in the curves E occurring for real values of h and k, and even if one is interested only in their real points, one has to take into account some non-real values of the parameter μ, those that correspond to two conjugated double roots for δ.

The (affine!) curve $\mathcal{E}$ has three real branches (Figure 15) but w' becomes infinite when w vanishes ($\mu = b_i^2$), so that these three branches must be considered as six half-branches, which will give six branches to the (real) discriminant. Figure 16 shows the real points of the discriminant (see also the close-up details in Figure 17) (recall that we have assumed that $b_0 = 0$).

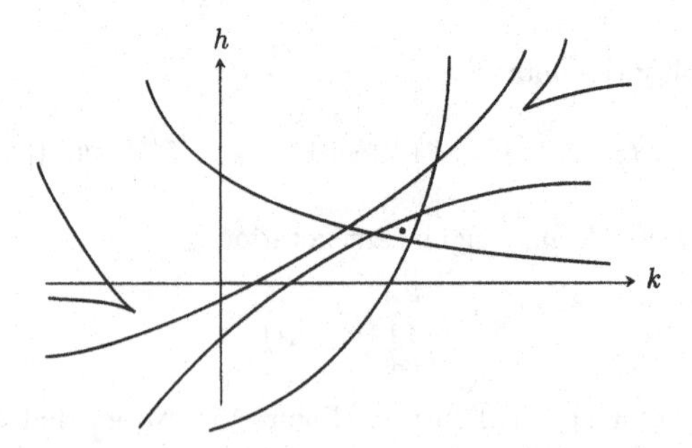

Figure 16: The real part of the discriminant.

If $f_2 = 0$, then C will be reducible to the union of a double line ($\lambda^2 = 0$) and a hyperelliptic curve

$$X: \qquad \lambda^2 = f_1 \frac{\mu^2 - h\mu + k}{\prod (b_i^2 - \mu)},$$

the double cover of $\mathbf{P}^1$ branched at the points b_i^2 and at the two roots μ_1 and μ_2 of the numerator, so that X has genus 2.

The discriminant of the family of curves X obtained when h and k vary is the union of the parabola $k = h^2/4$ (the numerator in the equation of X has a double root) and the four lines $k = hb_i^2 - b_i^4$ (the numerator and denominator have a common root) that are tangent to the parabola.

Remarks

1. These four lines already appear in the general picture ($f_2 \neq 0$), being the asymptotes of the discriminant.

2. This case ($f_2 = 0$) is the case where the orbit $\mathcal{O}_f$ consists of rank-2 skew-symmetric matrices. The system under consideration is then a "Moser system" [63], as is, for instance, the system giving the geodesics of an ellipsoid (see Knörrer's paper [52] and the author's papers [13, 11]).

3.3. The regular levels and the image of the momentum mapping

The method explained in Appendix 3 (A.3.4) and already used in II.2.3.4 and III.2.3.4 can be used here very simply to prove:

3.3.1 PROPOSITION. *If (h, k) is such that the genus-1 curve E (resp. the hyperelliptic curve X if $f_2 = 0$) is smooth, then it is a regular value of the momentum mapping.*

Proof. We use the proof of Proposition 3.1.1 and the computations of the tangent mapping to the eigenvector mapping explained in Appendix 3. As the eigenvalue corresponding to a point (λ, μ) of X is $\lambda\mu$, the subspace of $H^1(\mathcal{O}_C)$ generated by the images of the Hamiltonian vector fields of H and K is that generated by the cocycles $\lambda\mu^3$ and $\lambda\mu^2$ (as usual in this text, we are speaking of cocycles of the covering $C = \mathcal{U}_+ \cup \mathcal{U}_-$, as in Appendix 3).

It is worthwhile noting that the two cocycles are anti-invariant with respect to the involution $(\lambda, \mu) \mapsto (-\lambda, \mu)$. This remark will be useful in 3.4.

Let us check now that their classes are independent. To do this, we will compute their residue against some holomorphic forms on C. The four points P_0, P_1, P_2, P_3 are exactly the points of C where $\lambda = \infty$. In other words,

$$C - \mathcal{U}_+ = \{P_0, P_1, P_2, P_3\}.$$

This is why, for any holomorphic 1-form α on C, $[f] \mapsto \sum \operatorname{Res}_{P_i}(f\alpha)$ defines a linear form on $H^1(\mathcal{O}_C)$.

Let us put now $t = 1/z$, so that the equation for E becomes

$$t^2 + 2ta(\mu) + b(\mu) = 0$$

and it becomes obvious that

$$\omega = \frac{d\mu}{t + a(\mu)}$$

(resp. $\eta = p^\star\omega$) is a holomorphic 1-form on E (resp. on C). A simple computation of divisors shows that η, $\lambda\eta$ and $\lambda(b_0^2 - \mu)\eta$ are holomorphic and independent on C.

Put $u = 1/\lambda$ and remember that we have assumed $b_0 = 0$, so that the only thing left to do is to check that the two vectors

$$\left(\sum_{i=0}^{3} \operatorname{Res}_{P_i} \frac{\mu^2\eta}{u^2}, \quad \sum_{i=0}^{3} \operatorname{Res}_{P_i} \frac{\mu^3\eta}{u^2}\right)$$

and

$$\left(\sum_{i=0}^{3} \operatorname{Res}_{P_i} \frac{\mu^3\eta}{u^2}, \ \sum_{i=0}^{3} \operatorname{Res}_{P_i} \frac{\mu^4\eta}{u^2}\right)$$

are independent as vectors of $\mathbf{C}^2$. Now

$$\operatorname{Res}_{P_i} \frac{\mu^j\eta}{u^2} = \frac{4b_i^2}{j}$$

(by straightforward computation), so that our two vectors are actually

$$4\sum_{i=0}^{3} b_i^2 \left(\frac{1}{2}, \frac{1}{3}\right) \quad \text{and} \quad 4\sum_{i=0}^{3} b_i^2 \left(\frac{1}{3}, \frac{1}{4}\right),$$

a computation that ends the proof of 3.3.1. □

The critical values of (h, k) thus belong to the discriminant described above. As the orbit $\mathcal{O}_f$ is compact, the image of (h, k) is a compact subset of $\mathbf{R}^2$, and is thus contained in the shaded part of Figure 16, whose possible form according to the value of α is shown in Figure 17. I was helped by the software MAPLE to understand how to construct Figures 16, 17 and 18. The unshaded part of Figure 17 (and not a single square millimeter more) is also in Oshemkov's announcement [68] and in his paper in Fomenko's book [32].

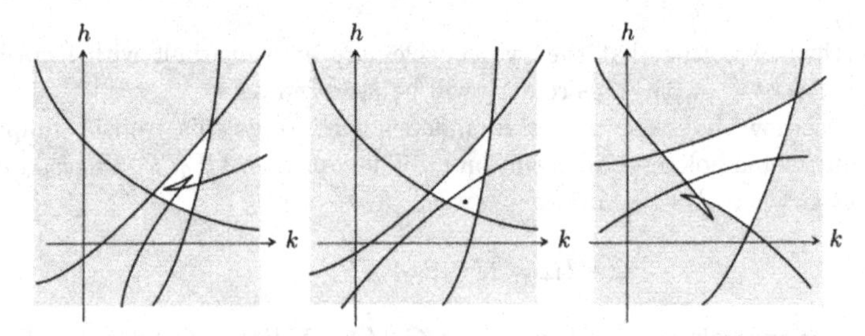

Figure 17

I leave to the interested reader the verification that the image of the momentum mapping is indeed the whole unshaded part of Figure 17 and that all the points of the discriminant belonging to this region are actually critical points: I suggest using the points $x_i = y_i = 0$, $x_j = y_j = 0$ ($\{i, j\} \subset \{1, 2, 3\}$) in $\mathcal{O}_f$.

3.4. The topology: another approach

As was noticed by Haine, the eigenvector mapping is a four-fold cover of its image:

3.4.1 PROPOSITION (Haine [39]). *If $f_2 \neq 0$ and if C is smooth, the eigenvector mapping*

$$\varphi_C : \mathcal{T}_{h,k}^{\mathbf{C}} \longrightarrow \operatorname{Pic}^6(C)$$

is a covering of group $\mathbf{Z}/2 \times \mathbf{Z}/2$ of its image. This image is isomorphic to an open subset of $\operatorname{Prym}(C|E)$.

Remark. Once again, the degree can be computed by the Riemann-Roch theorem as in Appendix 3 and in III.2.3.

After the proof of 3.3.1, it is clear that φ_C is a covering map. Its image must be a subvariety parallel to $\mathrm{Prym}(C|E)$, since we have noticed that the cocycles were anti-invariant. The matrices $M = (x, y)$ such that all the corresponding $M + J^2\lambda$ belong to the same conjugacy class modulo $GL(4;\mathbf{C})$ (*and* are in the same $SO(4)$-orbit) are given by

$$\left(\begin{pmatrix} \varepsilon x_1 \\ \eta x_2 \\ \varepsilon\eta x_3 \end{pmatrix}, \begin{pmatrix} \varepsilon y_1 \\ \eta y_2 \\ \varepsilon\eta y_3 \end{pmatrix}\right) \quad \text{with } \varepsilon^2 = \eta^2 = 1. \quad \square$$

The main result of Haine's paper [39] is that $\mathcal{T}^{\mathbf{C}}_{h,k}$ can be compactified by the addition of a curve to become an Abelian surface A *dual* to $\mathrm{Prym}(C|E)$.

Remark. It is possible to understand an extension $A \to \mathrm{Prym}(C|E)$ of the eigenvector mapping as the map from A to its dual given by the polarisation. Both Abelian varieties have a $(1, 2)$-type polarisation, and this explains why the kernel is a $\mathbf{Z}/2 \times \mathbf{Z}/2$.

Haine has also shown that A itself is the Prym variety of a covering of a genus-1 curve (precisely the curve $\mathcal{E}$ used above!) by a genus-3 curve:

$$D : \begin{cases} w^2 &= \prod (b_i^2 - \mu) \\ v^2 &= -\alpha w - \mu^2 + h\mu - k \end{cases}$$

(with the notation of 3.2). This duality is explained in Appendix 5.

All the desirable topological information can be deduced from this result. Indeed the real structures of D and $\mathcal{E}$ are very simple, as their equations show.

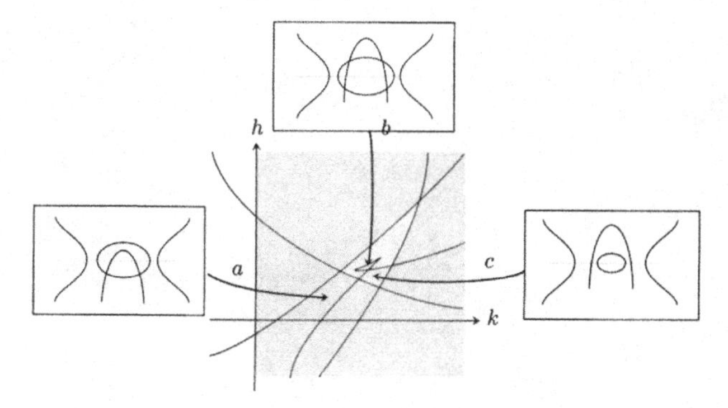

Figure 18

Figure 18 shows, according to the values of h and k, the relative positions of the parabola $\alpha w + \mu^2 - h\mu + k = 0$ and the curve $\mathcal{E}_{\mathbf{R}}$. The discriminant (Figure 16) corresponds to the values of h and k for which these two curves are tangent to each other (another exercise for the reader). The values of (μ, w) that give two real points of D are those that are both on $\mathcal{E}_{\mathbf{R}}$ and inside the parabola. Thus $D_{\mathbf{R}}$ has *one* component (over a part of one component of $\mathcal{E}_{\mathbf{R}}$) in the region labelled a, two (each over the same component of $\mathcal{E}_{\mathbf{R}}$) in the region b and two over a component of $\mathcal{E}_{\mathbf{R}}$ in the region c. The structure of $\text{Prym}(D|\mathcal{E})_{\mathbf{R}}$ can now be deduced in the same way as for the Kowalevski top (Chapter III and Audin & Silhol [15]) and thus also the Liouville tori and their bifurcations. I have no doubt that this would give the results announced by Oshemkov [68].

V

Non-compact levels: a Toda lattice

In this chapter, I will describe an avatar of the periodic Toda lattice, chosen especially because it has a non-proper Hamiltonian, non-complete flows and non-compact level sets. The aim is to show that the properties of the eigenvectors of the Lax matrices for this system (the classical results of van Moerbeke & Mumford [62]) can be used to understand the topology of the level sets even if they are non-compact.

1. The differential system and the spectral curve

1.1. The periodic Toda lattice

The periodic Toda lattice is the differential system described by the Lax equation

$$\dot{A_\lambda} = [A_\lambda, B_\lambda] \tag{1}$$

with

$$A_\lambda = \begin{pmatrix} b_1 & a_1 & 0 & & & a_{n+1}\lambda^{-1} \\ a_1 & b_2 & a_2 & & & \\ 0 & a_2 & \ddots & \ddots & & \\ & & \ddots & \ddots & \ddots & 0 \\ & & & \ddots & \ddots & a_n \\ a_{n+1}\lambda & & & 0 & a_n & b_{n+1} \end{pmatrix},$$

$$B_\lambda = \begin{pmatrix} 0 & a_1 & 0 & & & -a_{n+1}\lambda^{-1} \\ -a_1 & 0 & a_2 & & & \\ 0 & -a_2 & \ddots & \ddots & & \\ & & \ddots & \ddots & \ddots & 0 \\ & & & \ddots & \ddots & a_n \\ a_{n+1}\lambda & & & 0 & -a_n & 0 \end{pmatrix},$$

that is,

$$(2)\qquad \begin{cases} \dot{a_k} = a_k(b_k - b_{k+1}) & (b_{n+2} = b_1) \\ \dot{b_k} = 2(a_{k-1}^2 - a_k^2) & (a_0 = a_{n+1}). \end{cases}$$

Notice that $\sum \dot{a_k}/a_k = 0$ and $\sum \dot{b_k} = 0$ so that $\prod a_k$ and $\sum b_k$ are obvious first integrals.

Remark. This is supposed to describe a system of $n+1$ particles connected by so-called "exponential springs", with a periodicity assumption: one can imagine that the particles are moving on a circle (see the details and the celebrated Flaschka change of coordinates in [28]).

Now, it is well known[1] that the Lax equation (1) describes a completely integrable system. I will give here a brief outline of a proof (sending the reader to Adler & van Moerbeke [2] for the details). Notice that in the differential system (2) the $n+1$ particles play analogous roles, but that the $(n+1)$th seems to have a specific role in the Lax matrix. Multiplication by λ looks very much like a shift in the indices, so that one is tempted to consider A_λ and B_λ as infinite periodic tridiagonal (Jacobi) matrices

$$A = \begin{pmatrix} \ddots & \ddots & \ddots & & & & & \\ a_{n+1} & b_1 & a_1 & 0 & & & & \\ \ddots & a_1 & b_2 & a_2 & & & & \\ & 0 & a_2 & \ddots & \ddots & & & \\ & & & \ddots & \ddots & \ddots & 0 & \\ & & & & \ddots & \ddots & a_n & \ddots \\ & & & & 0 & a_n & b_{n+1} & a_{n+1} \\ & & & & & \ddots & \ddots & \ddots \end{pmatrix}$$

and

$$B = \begin{pmatrix} \ddots & \ddots & \ddots & & & & & \\ -a_{n+1} & 0 & a_1 & 0 & & & & \\ \ddots & -a_1 & 0 & a_2 & & & & \\ & 0 & -a_2 & \ddots & \ddots & & & \\ & & & \ddots & \ddots & \ddots & 0 & \\ & & & & \ddots & \ddots & a_n & \\ & & & & 0 & -a_n & 0 & a_{n+1} \\ & & & & & \ddots & \ddots & \ddots \end{pmatrix}.$$

The first point is that the relevant infinite-dimensional analogue of $\mathfrak{sl}_{n+1}$ that we are now in is a Lie algebra isomorphic with $\mathfrak{sl}_{n+1}[\lambda, \lambda^{-1}]$, so that our differential system is indeed

$$\dot{A} = [A, B].$$

[1] There is a multitude of papers on the Toda lattice, among which are several of interest for the problems we address here.

Notice that now A is a symmetric matrix while B is skew-symmetric: the situation is precisely the one described in Appendix 2 (A.2.3), except that the dimension has become infinite. Now, and this is the second point, the AKS theorem could be applied here to prove that the system is actually integrable: but this is the place where the present discussion will be left incomplete. All we need to do is to check that everything works in our new infinite-dimensional framework. Once this is agreed, the system is seen to be the Hamiltonian system for the function

$$H = \frac{1}{2}\operatorname{tr} A_\lambda^2 = \frac{1}{2}\sum_{k=1}^{n+1} b_k^2 + \sum_{k=1}^{n+1} a_k^2$$

on the set of all tridiagonal symmetric (infinite, periodic, traceless) matrices, which is a union of coadjoint orbits and thus a Poisson manifold. All the coefficients of the polynomials $\operatorname{tr} A_\lambda^k$ are commuting first integrals.

Symplectic orbits. From the group-theoretical approach, it is obvious that the trace $\sum b_i$ is an orbit invariant. The same is also true of the product $\prod a_i$.

Once we have agreed that everything works in the infinite-dimensional framework, the previous statement is not very hard to prove. In the formulation (and the notation) of Appendix 2, with the invariant symmetric bilinear form

$$\left\langle \sum x_i\lambda^i, \sum y_j\lambda^j \right\rangle = \sum_{i+j=0} x_i \cdot y_j$$

(using the Killing form), for $x = \sum x_i\lambda^i$ we have

$$(\nabla_x f)_+ = \sum (\nabla_{x_i} f)\,\lambda^{-i}$$

(recall that $_+$ denotes projection on the Lie algebra $\mathfrak{a}$ of lower-triangular matrices) so that the Poisson bracket in $\mathfrak{b}^\perp = \mathfrak{a}^\star$ is

$$\begin{aligned} \{f, g\}(x) &= \left\langle x, \left[(\nabla_x f)_+, (\nabla_x g)_+\right]\right\rangle \\ &= \sum_{\substack{i-(j+k)=0 \\ j,k\geq 0}} \left\langle x_i, \left[\nabla_{x_j} f, \nabla_{x_k} g\right]\right\rangle. \end{aligned}$$

Our set Γ of tridiagonal matrices consists of elements of the form

$${}^t x_1\lambda^{-1} + x_0 + x_1\lambda$$

on which the bracket has the simpler expression

$$\{f, g\}(x) = \langle x_0, [\nabla_{x_0} f, \nabla_{x_0} g]\rangle + \langle x_1, [\nabla_{x_1} f, \nabla_{x_0} g] + [\nabla_{x_0} f, \nabla_{x_1} g]\rangle.$$

Very concretely, for a matrix A_λ as above, x_0 is the "constant term" and x_1 is the matrix whose unique possibly-non-zero entry is the a_{n+1} in the lower left-hand corner, so that, eventually

$$\{f, g\}(a, b) = \sum_{i=1}^{n+1} a_i \frac{\partial g}{\partial a_i}\left(\frac{\partial f}{\partial b_i} - \frac{\partial f}{\partial b_{i+1}}\right) + \sum_{i=1}^{n+1} \frac{\partial g}{\partial b_i}\left(a_{i-1}\frac{\partial f}{\partial a_i} - a_i\frac{\partial f}{\partial a_{i-1}}\right),$$

with obvious periodic notation for the indices. The Hamiltonian vector field associated with the function g can thus be expressed as

$$X_g = \sum_i \left(a_i \frac{\partial g}{\partial a_i} - a_{i-1} \frac{\partial g}{\partial a_{i-1}} \right) \frac{\partial}{\partial b_i} + \sum_i \left(a_{i-1} \frac{\partial g}{\partial b_i} - a_{i+1} \frac{\partial g}{\partial b_{i+1}} \right) \frac{\partial}{\partial a_i}.$$

From this, we see immediately that for $f = \prod a_i$ we have $X_f \equiv 0$, so that f is indeed an orbit invariant. One can even compute the rank of the Poisson structure from this formula and check that

$$V = \left\{ (a, b) \in \mathbf{R}^{n+1} \times \mathbf{R}^{n+1} \mid \sum b_i = 0 \text{ and } \prod a_i = 1 \right\}$$

is a symplectic leaf. We shall restrict ourselves to this manifold.

1.2. The spectral curve and the regular levels

As we are on V, the equation of the spectral curve is

$$\lambda + \lambda^{-1} + P(\mu) = 0$$

where P is the degree-$(n+1)$ polynomial

$$P(\mu) = (-1)^{n+1} \mu^{n+1} + H_1(b, a) \mu^{n-1} + \cdots + H_n(b, a).$$

The spectral curve can be completed (and normalised) by the addition of two points A and B such that the completed curve is smooth around A and B and

$$(\lambda) = (n+1)(A - B), \qquad (\mu) + A + B \geq 0.$$

Roughly speaking, in (λ, μ) coordinates, $A = (0, \infty)$ and $B = (\infty, \infty)$.

The resulting complete curve X is a genus-n hyperelliptic curve. The hyperelliptic involution τ is $(\lambda, \mu) \mapsto (\lambda^{-1}, \mu)$, the degree-2 map is μ and it is branched at the $2n + 2$ roots of $P(\mu)^2 - 4$.

Remark. The feature that is very special in this situation is that we are given on the curve X a divisor $A - B = A - \tau(A)$ representing an element of order $n + 1$ in $\mathrm{Pic}^0(X)$. In particular, this element defines a $\mathbf{Z}/(n+1)$-action on any $\mathrm{Pic}^d(X)$.

1.2.1 PROPOSITION. *The eigenvector mapping*

$$\varphi : \mathcal{T}_h \longrightarrow \mathrm{Pic}^d(X)$$

is equivariant with respect to $\mathbf{Z}/(n+1)$-actions, that is,

$$\varphi\sigma(b, x) = \varphi(b, x) + A - B.$$

Proof. Obviously,

$$A_\lambda \begin{pmatrix} u_1 \\ \vdots \\ u_n \\ u_{n+1} \end{pmatrix} = \mu \begin{pmatrix} u_1 \\ \vdots \\ u_n \\ u_{n+1} \end{pmatrix} \iff \sigma A_\lambda \begin{pmatrix} \lambda^{-1} u_{n+1} \\ u_1 \\ \vdots \\ u_n \end{pmatrix} = \mu \begin{pmatrix} \lambda^{-1} u_{n+1} \\ u_1 \\ \vdots \\ u_n \end{pmatrix}$$

so that $\varphi(A_\lambda)$ is the "poles − zeros" divisor for the section $u = {}^t(u_1, \ldots, u_{n+1})$ of the eigenvector bundle if and only if $\varphi(\sigma(A_\lambda))$ is the "poles − zeros" divisor for the section ${}^t(\lambda^{-1} u_{n+1}, u_1, \ldots, u_n)$. It is now easily checked that the latter is obtained from the former by adding $A - B$. □

Notice that, as usual, the map φ is well-defined once the curve X is smooth. Moreover we have:

1.2.2 PROPOSITION. *Assume the value h is such that the corresponding spectral curve X is smooth. Then $\mathcal{T}_h$ is a regular level.*

Proof. To prove that h is a regular value, it is enough to exhibit n vector fields $X_1, \ldots, X_n$ that are tangent to $\mathcal{T}_h$ and independent at all points of $\mathcal{T}_h$, in other words n skew-symmetric matrices $B_\lambda^{(1)}, \ldots, B_\lambda^{(n)}$ such that the images of the brackets $\left[A_\lambda, B_\lambda^{(k)}\right]$ by the tangent mapping to φ are linearly independent. The easiest thing to do is to take for $B_\lambda^{(k)}$ the skew-symmetric part of $(A_\lambda)^k$ (so that $B_\lambda^{(1)}$ is our actual matrix B_λ). Suppose now that v is an eigenvector of A_λ with respect to the eigenvalue μ,

$$\mu^k v = (A_\lambda)^k v = B_\lambda^{(k)} v + \text{terms involving only } \lambda^{-1}.$$

From Appendix 3, one deduces that the image of the vector field $\left[A_\lambda, B_\lambda^{(k)}\right]$ is μ^k, viewed as an element of $H^1(X; \mathcal{O}_X)$. The only thing to check now is that $\mu, \ldots, \mu^n$ form a basis of this n-dimensional vector space; this is easily done by a residue computation:

1.2.3 LEMMA. *Write $y = 2\lambda + P(\mu)$, so that the equation for X becomes $y^2 = P(\mu)^2 - 4$. Then $\mu^n, \ldots, \mu$ is the basis of $H^1(X; \mathcal{O}_X)$ that is dual to the basis $d\mu/y, \ldots, \mu^{n-1} d\mu/y$ of $H^0(X; \Omega_X^1)$.* □

1.3. The Belgian change of coordinates

Recall that we are looking for an example with a non-proper Hamiltonian. The easiest thing one can try in order to find one is to replace the Hamiltonian by something like

$$H = \frac{1}{2} \sum_{k=1}^{n+1} b_k^2 + \sum_{k=1}^{n+1} x_k$$

in other words to perform the "change of variables" $x_k = a_k^2$ and to arrive in the symplectic manifold

$$W = \left\{(x, b) \in \mathbf{R}^{n+1} \times \mathbf{R}^{n+1} \mid \sum b_i = 0 \text{ and } \prod x_i = 1\right\}$$

so that the Hamiltonian system is

$$\text{(3)} \qquad \left\{ \begin{array}{rcl} \dot{x}_k & = & 2x_k(b_k - b_{k+1}) \\ \dot{b}_k & = & 2(x_{k-1} - x_k). \end{array} \right.$$

The reason why this can work is that the first integrals for system (2) depend only on the $a_k^2 = x_k$. This is so because, if D is the diagonal matrix

$$D = (a_1 \ldots a_{n+1}, a_2 \ldots a_{n+1}, \ldots, a_n a_{n+1}, a_{n+1}),$$

then the matrix $D^{-1} A_\lambda D$ depends only on the squares x_k, so that the same is true of the coefficients in the characteristic polynomial. System (3) is thus completely integrable.

More precisely, the commuting integrals are given by $\operatorname{tr}(A'_\lambda)^k$, where

$$A'_\lambda = D^{-1}A_\lambda D = \begin{pmatrix} b_1 & 1 & 0 & & & x_{n+1}\lambda^{-1} \\ x_1 & b_2 & 1 & & & \\ 0 & x_2 & \ddots & \ddots & & \\ & & \ddots & \ddots & \ddots & 0 \\ & & & \ddots & \ddots & 1 \\ \lambda & & & 0 & x_n & b_{n+1} \end{pmatrix}.$$

These are, of course, exactly the first integrals we had before but with a_k^2 replaced by x_k. I shall thus write $H_i(b,x)$ instead of $H_i(b,a)$. The spectral curve X has not changed during this operation.

Remark. System (3) is also equivalent to a Lax equation for A'_λ. Directly, we can write

$$\frac{d}{dt}A'_\lambda = \left(\frac{d}{dt}D^{-1}\right)A_\lambda D + D^{-1}\left(\frac{d}{dt}A_\lambda\right)D + D^{-1}A_\lambda\left(\frac{d}{dt}D\right)$$

so that

$$\frac{d}{dt}A'_\lambda = [A'_\lambda, C_\lambda]$$

with

$$C_\lambda = D^{-1}B_\lambda D + D^{-1}\frac{d}{dt}D.$$

Now, the diagonal matrix $D^{-1}dD/dt$ has $(a_1\dots a_{k-1})d(a_k\dots a_{n+1})/dt$ as kth entry, which gives, since $\prod a_i = 1$,

$$\frac{d}{dt}\log(a_k\dots a_{n+1}) = \left(\frac{\dot a_k}{a_k} + \cdots + \frac{\dot a_{n+1}}{a_{n+1}}\right) = b_k - b_1$$

so that

$$C_\lambda = \begin{pmatrix} b_1 & 1 & 0 & & & -x_{n+1}\lambda^{-1} \\ -x_1 & b_2 & 1 & & & \\ 0 & -x_2 & \ddots & \ddots & & \\ & & \ddots & \ddots & \ddots & 0 \\ & & & \ddots & \ddots & 1 \\ \lambda & & & 0 & -x_n & b_{n+1} \end{pmatrix} - b_1\,\mathrm{Id},$$

and

$$C_\lambda + b_1\,\mathrm{Id} - A'_\lambda = \begin{pmatrix} 0 & 0 & 0 & & & -2x_{n+1}\lambda^{-1} \\ -2x_1 & 0 & 0 & & & \\ 0 & -2x_2 & \ddots & \ddots & & \\ & & \ddots & \ddots & \ddots & 0 \\ & & & \ddots & \ddots & 0 \\ 0 & & & 0 & -2x_n & 0 \end{pmatrix} = B'_\lambda$$

and our system is eventually equivalent to

$$\frac{d}{dt}A'_\lambda = [A'_\lambda, B'_\lambda].$$

There are of course group-theoretical reasons for this Lax form (see e.g. Kostant's extended paper [54]).

Let us denote by $\mathcal{T}'_h$ the level h in W and by φ' the new eigenvector mapping. Note that, at least from the complex point of view, $\mathcal{T}_h$ is a $(\mathbf{Z}/2)^n$-fold cover of $\mathcal{T}'_h$. We have a commutative diagram

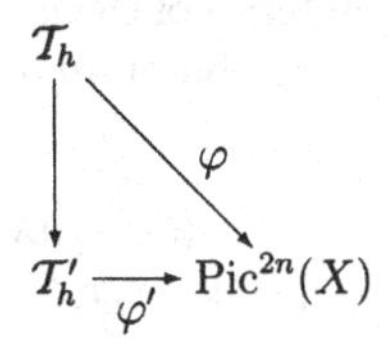

and the cyclic permutation σ still acts on $\mathcal{T}'_h$, with commutative diagrams

$$\begin{array}{ccc} \mathcal{T}_h & \xrightarrow{\sigma} & \mathcal{T}_h \\ \downarrow & & \downarrow \\ \mathcal{T}'_h & \xrightarrow{\sigma} & \mathcal{T}'_h \end{array} \qquad \text{and} \qquad \begin{array}{ccc} \mathcal{T}'_h & \xrightarrow{\varphi'} & \mathrm{Pic}^{2n}(X) \\ \sigma\downarrow & & \downarrow A-B \\ \mathcal{T}'_h & \xrightarrow{\varphi'} & \mathrm{Pic}^{2n}(X) \end{array}$$

The explicit computations of van Moerbeke & Mumford [62] allow us to exhibit an effective divisor D, of degree n, such that $D + nA$ is a representative of $\varphi(A'_\lambda)$. The divisor D is *general* and satisfies

$$h^0(D + (n-k+1)A - (n-k+1)B) = 1 \qquad \forall k \in \{1, \dots, n\}$$

which makes it possible to determine very precisely the image of the map φ.

All this is very classical (see van Moerbeke & Mumford [62], Adler & van Moerbeke [2, 6] and the author's paper [12]). We will content ourselves here with the $n = 2$ case, not because the arguments are simpler (they are actually the same) but because this is the case for which we want to derive a topological description of the level sets $\mathcal{T}'_h$.

2. The eigenvector mapping: the $n = 2$ case

We look now at 3×3 matrices (that is, at three particles).

2.1. The curve and the integrals

The hyperelliptic curve has genus 2 and is a completion of the curve

$$X_0 = \left\{(\lambda, \mu) \in \mathbf{C}^\star \times \mathbf{C} \mid \lambda + \lambda^{-1} - \mu^3 + H_1(b,x)\mu + H_2(b,x) = 0\right\}$$

where

$$\begin{cases} H_1(b,x) &= x_1 + x_2 + x_3 - b_1b_2 - b_2b_3 - b_3b_1 \\ H_2(b,x) &= b_1b_2b_3 - b_1x_2 - b_2x_3 - b_3x_1. \end{cases}$$

Fix values h_1, h_2 of H_1 and H_2. The corresponding curve X is smooth when the roots of

$$\Delta(\mu) = \left(-\mu^3 + h_1\mu + h_2\right)^2 - 4$$

are distinct, that is, when the two polynomials $-\mu^3 + h_1\mu + h_2 \pm 2$ have simple roots.

2.2. Properties of the eigenvector mapping

The eigenvectors are easily computed in terms of the minors of the matrix $A'_\lambda - \mu \operatorname{Id}$ (note that the computation is quite general and comes from van Moerbeke & Mumford [62]). It is found that

$$\begin{pmatrix} u_1 \\ u_2 \\ u_3 \end{pmatrix} = \begin{pmatrix} -1 + x_3(b_2 - \mu)\lambda^{-1} \\ b_1 - \mu - x_1x_3\lambda^{-1} \\ x_1 - (b_1 - \mu)(b_2 - \mu) \end{pmatrix}$$

is an eigenvector of A'_λ for the eigenvalue μ.

Notice that u_3 is a function of μ so that it vanishes at two pairs of points $(Q_1, \tau Q_1)$, $(Q_2, \tau Q_2)$. Let us decide that Q_1 and Q_2 are the common zeros of u_1 and u_3, and note that $u_1 = u_3 = 0 \Rightarrow u_2 = 0$ so that they are also the common zeros of u_2 and u_3.

Set $P_i = \tau Q_i$ and look at the section ${}^t(f_1, f_2, 1) = {}^t\left(\frac{u_1}{u_2}, \frac{u_2}{u_3}, 1\right)$ of the eigenvector bundle.

Note that the order of the pole at A is:

- 2 in μ^2 and u_3
- 3 in λ and u_2
- 4 in $\lambda\mu$ and u_1

and similarly that B is not a pole of f_1 and f_2, so that

$$\begin{aligned} (f_1)_\infty &= P_1 + P_2 + 2A \\ (f_2)_\infty &= P_1 + P_2 + A \end{aligned}$$

and the divisor of poles of the section is $P_1 + P_2 + 2A$. Notice similarly that B is a simple zero of f_2 and a double zero of f_1.

In this way we have obtained a representative of the image of (b, x) in $\operatorname{Pic}^4(X)$ via the eigenvector mapping. The divisor D mentioned above is nothing other than $P_1 + P_2$. The next aim is to look at its properties.

2.2.1 PROPOSITION. *There exists an effective divisor D of degree 2 on X which is general, which satisfies*

$$h^0(D + kA - (k+1)B) = 0 \qquad \textit{for} \quad k = 0, 1, 2$$

and which is such that $D + 2A$ is a representative of the eigenvector divisor class.

Proof. First of all, for $m \geq 1$, $\deg(D + mA) = m + 2 > 2$ so that, according to the Riemann-Roch theorem,

$$h^0(D + mA) = m + 2 - 2 + 1 = m + 1.$$

For instance $h^0(D + A) = 2$. But this implies that $h^0(D) = 1$ since we have noticed that

$$f_2 \in \mathcal{L}(D + A) - \mathcal{L}(D)$$

so that the inclusion is strict. That is to say, D is *general*. As a consequence, $\mathcal{L}(D) = \mathbf{C}$ (constant functions) so that $\mathcal{L}(D - B) = 0$ (the constant is zero) and, since $f_2 \in \mathcal{L}(D + A - B)$, $h^0(D + A - B) = 1$ (recall that to add a pole can at most increase the dimension by 1!).

Also, the zero of f_2 at A is simple, so that

$$f_2 \in \mathcal{L}(D + A - B) - \mathcal{L}(D + A - 2B)$$

and $h^0(D + A - 2B) = 0$. Similarly, $f_1 \in \mathcal{L}(D + 2A - 2B)$, so that $h^0(D + 2A - 2B) = h^0(D + A - 2B) + 1 = 1$ and $h^0(D + 2A - 3B) = 0$. □

Suppose conversely that we are given a divisor D, on the curve X, that satisfies the above properties. It is possible to reconstruct a matrix A'_λ of which $D + 2A$ is the image. Thus we will get our main result.

Consider the Abel-Jacobi mapping

$$\begin{array}{ccc} X & \longrightarrow & \mathrm{Pic}^4(X) \\ P & \longmapsto & P + 3A \end{array}$$

and call $\mathcal{D}_1$ its image (this is a copy of X embedded in $\mathrm{Pic}^4(X)$, a translate of the Θ-divisor). Similarly, call $\mathcal{D}_2$ and $\mathcal{D}_3$ the images of $\mathcal{D}_1$ resulting from order-3 translations by $A - B$ and $2A - 2B$ respectively.

2.2.2 THEOREM. *Assume that the curve X is smooth. Then $\mathcal{T}'_h$ is a regular level, the eigenvector mapping*

$$\varphi' : \mathcal{T}'_h \longrightarrow \mathrm{Pic}^4(X)$$

is an isomorphism onto its image and the image of φ' is the complement of the union $\mathcal{D}_1 \cup \mathcal{D}_2 \cup \mathcal{D}_3$.

Proof. The first assertion follows readily from Proposition 1.2.2. Let us now prove the second assertion, that is, try to reconstruct a matrix A'_λ from a divisor D. First of all, $\mathcal{L}(D + 2B) = V$ is a three-dimensional vector space: this will be the $\mathbf{C}^3$ on which the matrix A'_λ will act. Moreover, we can choose a basis $(g_1, g_2, 1)$ of V such that

$$g_1 \in \mathcal{L}(D + 2A - 2B), \qquad g_2 \in \mathcal{L}(D + A - B).$$

As both subspaces of $\mathcal{L}(D+2A)$ are one-dimensional, g_1 and g_2 are well-defined up to multiplication by a non-zero complex number. What we need to understand is multiplication by μ on these functions.

Consider first μg_2.

$$(\mu g_2) = (\mu) + (g_2) \geq -B - A - D - A + B = -D - 2A$$

so that $\mu g_2 \in \mathcal{L}(D + 2A)$ and

$$\mu g_2 = y_1 g_1 + b_2 g_2 + c.$$

Also, since we know that $\mathcal{L}(D + A - 2B) = 0$, the zero of g_2 at B is simple, so that μg_2 cannot vanish at B and $c \neq 0$. Now put $f_2 = g_2/c$ so that

$$\mu f_2 = y_1 g_1 + b_2 f_2 + 1.$$

Look now at μg_1:

$$(\mu g_1) = (\mu) + (g_1) \geq -B - A + (g_1) \geq -B - A - D - 2A + 2B.$$

Thus $(\mu g_1) \geq -D + B - 3A$: the meromorphic function μg_1 has poles on D and a triple pole at A but no other pole; it vanishes at B. Recall that λ also has a triple pole at A, so that there exists a non-zero complex number y_3 such that $\mu g_1 - y_3\lambda^{-1}$ has, at worst, a pole of order 2 at A. Thus $\mu g_1 - y_3\lambda^{-1} \in \mathcal{L}(D + 2A - B)$. Now, (g_1, f_2) is a basis of $\mathcal{L}(D + 2A - B)$ so that we can write

$$\mu g_1 - y_3\lambda^{-1} = b_1 g_1 + c f_2.$$

Lastly, look at $\mu \cdot 1$. Recall that at B, the function μ has a simple pole and λ has a triple pole while g_1 has a double zero, so that λg_1 has a simple pole there. Replacing g_1 by a suitable multiple f_1, we can arrange that $\mu - \lambda f_1$ has no pole at B. Now, at A, λf_1 has a simple zero so that $\mu - \lambda f_1$ has a simple pole. Of course, λf_1 has simple poles at D and no other pole, so that $\mu - \lambda f_1 \in \mathcal{L}(D + A)$ and thus can be written $x_2 f_2 + b_3$ for well-defined scalars x_2 and b_3. Note that now f_1 and f_2 are completely determined and that they satisfy

$$\mu \begin{pmatrix} f_1 \\ f_2 \\ 1 \end{pmatrix} = \begin{pmatrix} b_1 & c' & x_3\lambda^{-1} \\ x_1 & b_2 & 1 \\ \lambda & x_2 & b_3 \end{pmatrix} \begin{pmatrix} f_1 \\ f_2 \\ 1 \end{pmatrix}.$$

As $(\lambda, \mu) \in X$, $c' = 1$ and we have reconstructed the matrix A'_λ from the divisor D.

Thus we have shown that any general divisor D satisfying the additional properties stated in Proposition 2.2.1 is the image of exactly one matrix A'_λ so that the eigenvector mapping is injective and the above properties describe its image.

The only thing to do now is to check that the properties listed in Proposition 2.2.1 mean exactly that $D + 2A \notin \mathcal{D}_i$. □

2.3. The topology of the level sets

We shall now derive the topology of the level sets T'_h from Theorem 2.2.2: they are modelled on the real part of the complement[2] of the divisors $\mathcal{D}_i$ in the Jacobian. We shall first look at the relative positions of the divisors $\mathcal{D}_i$ and then look at the way in which they are embedded in the Jacobian.

[2]Notice that this is the only example in this book where we need to know the image of the eigenvector mapping: in the case where the level under consideration is compact, once you have managed to obtain an injective eigenvector mapping you are done, your level being identified with the real part of the Abelian variety.

Relative positions of $\mathcal{D}_1$, $\mathcal{D}_2$ and $\mathcal{D}_3$. Each $\mathcal{D}_i$ is tangent to the following one. This is usually described by a combinatorial picture, as in Figure 19; a more accurate description appears in Figure 21. Let us state this property more precisely.

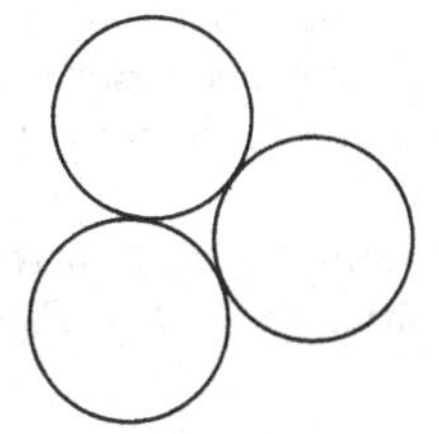

Figure 19

2.3.1 PROPOSITION. *The intersection of $\mathcal{D}_i$ and $\mathcal{D}_{i+1}$ (i mod 3) consists of one point, which is real, and at which the two curves are tangent. The tangent vector at the intersection point is the line generated by the cocycle $\mu \in H^1(X; \mathcal{O}_X)$.*

Proof. The homological intersection is, in general, given by the Poincaré formula (see e.g. Lange & Birkenhacke [57] for that kind of result). Here this is quite easy to find, since we are just dealing with a genus-2 curve embedded in a 4-torus: the self-intersection number is the Euler number of the normal bundle; since the ambient space is a torus, this is just the opposite of the Euler characteristic, thus equal to 2.

Now the set-theoretic intersection $\mathcal{D}_1 \cap \mathcal{D}_2$ contains the class of the divisor $4A$, which appears in $\mathcal{D}_1$ as the image $A + 3A$ of A and in $\mathcal{D}_2$ as the image $B + 3A + (A - B)$ of B. Note that $4A$ is a real point.

To prove the proposition, it is enough to prove that the line directed by the cocycle μ is tangent to both $\mathcal{D}_1$ and $\mathcal{D}_2$ at this point: this will ensure that $\mathcal{D}_1$ and $\mathcal{D}_2$ are tangent at this point, but, as the self-intersection number is 2, the point $4A$ will be the only point in $\mathcal{D}_1 \cap \mathcal{D}_2$.

Now, using the translations, what we aim to prove is that, in the image of X in its Jacobian via the Abel-Jacobi map, the tangent line at both points A and B is the line directed by the cocycle μ. This is fairly easy: write the Abel-Jacobi map as

$$\begin{array}{ccc} X & \longrightarrow & H^0\left(\Omega^1_X\right)^* / \Lambda \\ P & \longmapsto & \left(\omega \mapsto \int^P \omega\right) \end{array}$$

so that its tangent map is

$$\begin{array}{ccc} T_P X & \longrightarrow & H^0\left(\Omega^1_X\right) \\ \eta & \longmapsto & \left(\omega \mapsto \omega_P(\eta)\right) \end{array}$$

and the line tangent to the image curve at the point image of P is the line generated by the linear form $\omega \mapsto \omega_P$.

What we want to prove is that the lines in $H^0\left(\Omega^1_X\right)^*$ that are generated by $\omega \mapsto \omega_A$ and $\omega \mapsto \omega_B$ both coincide with that generated by the cocycle μ. Now, at both points A and B, $u = 1/\mu$ is a holomorphic local coordinate so that, if $\omega = \alpha(u)du$ at A and $\omega = \beta(u)du$ at B, $\mathrm{Res}_A(\mu\omega) = \alpha(0)$, $\mathrm{Res}_B(\mu\omega) = \beta(0)$ and

$$\begin{cases} \omega_A = 0 \Leftrightarrow \mathrm{Res}_A(\mu\omega) = 0 \\ \omega_B = 0 \Leftrightarrow \mathrm{Res}_B(\mu\omega) = 0 \end{cases}$$

which is precisely what we wanted to check. □

Remark. In the general case, we obtain $n+1$ translates of a divisor $\mathcal{D}_1$, each tangent to both the previous and the following divisors. The combinatorial picture reproduces the extended Dynkin diagram of the simple Lie algebra $\mathfrak{sl}_{n+1}$. This comes from Adler & van Moerbeke [6] (analogous results concerning Toda lattices associated with other simple Lie algebras can be found in the same paper). Both the statement and the proof given here are quite general (including the assertion on the tangent, see [12]).

The regular values. Let us now draw the real curve $X_{\mathbf{R}}$ (in (μ, y) coordinates as in Lemma 1.2.3) according to the position of (h_1, h_2) (Figure 20).

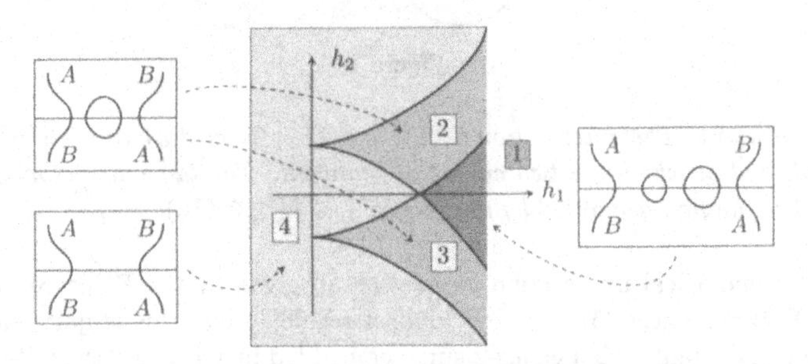

Figure 20

The discriminant curve consists of two copies of the discriminant of degree-3 polynomials (see 2.1). Outside these curves, we have regular levels, and one can check that the points of the discriminant are actually critical values. There are four regions in the complement of this discriminant, respectively labelled (1), (2), (3) and (4), which correspond to a real part $X_{\mathbf{R}}$ having respectively three, two, two or one connected component(s).

The real curve in its Jacobian. Notice that, as $X_{\mathbf{R}}$ is always non-empty here, $\text{Pic}^4(X)_{\mathbf{R}}$ may be identified with the real part of the Jacobian. The only thing we need now is to understand the real aspects of the Abel-Jacobi map.

2.3.2 LEMMA. *Any connected component of $X_{\mathbf{R}}$ represents a primitive non-zero element of $H_1(\text{Jac}(X)_{\mathbf{R}}; \mathbf{Z})$.*

Proof. Let X_0 be a connected component of $X_{\mathbf{R}}$, and let T_0 be the 2-torus to which it is sent by the Abel-Jacobi map u. Recall that $u_\star : H_1(X; \mathbf{Z}) \to H_1(\text{Jac}(X); \mathbf{Z})$ is an isomorphism (see Appendix 4).

If one thinks of X as of a double covering of $\mathbf{P}^1(\mathbf{C})$ branched at the roots of $P(\mu)^2 - 4$, it is more or less obvious that each component of the real part gives a primitive non-zero element in $H_1(X; \mathbf{Z})$ and thus in $H_1(\text{Jac}(X); \mathbf{Z})$.

Now a connected component T_0 of $\text{Jac}(X)_{\mathbf{R}}$ is defined by the inclusion $P \hookrightarrow \mathbf{C}^2$ of some translate of the real plane $\mathbf{R}^2$ and the corresponding lattice $\Lambda_0 \hookrightarrow \Lambda$, so that we have the commutative diagram

$$\begin{array}{ccc} \Lambda_0 = H_1(T_0) & \hookrightarrow & H_1(\text{Jac}(X)) = \Lambda \\ \uparrow & & \uparrow \\ H_1(X_0) & \hookrightarrow & H_1(X) \end{array}$$

and the generator of $H_1(X_0)$ is mapped onto a primitive non-zero element in $H_1(T_0)$. □

The complement of any component of $X_{\mathbf{R}}$ in its own torus is thus a cylinder. Using the relative positions of the divisors $\mathcal{D}_i$ as depicted in Proposition 2.3.1, one derives easily a complete description of the topology of the real level set $\mathcal{T}_h'$ as shown in Figure 21.

2.3.3 PROPOSITION. *The real level set $\mathcal{T}_h'$ consists of three open discs if $X_{\mathbf{R}}$ is connected, three open discs and three cylinders if $X_{\mathbf{R}}$ has two components, and three open discs, six cylinders and a torus if $X_{\mathbf{R}}$ has three connected components.* □

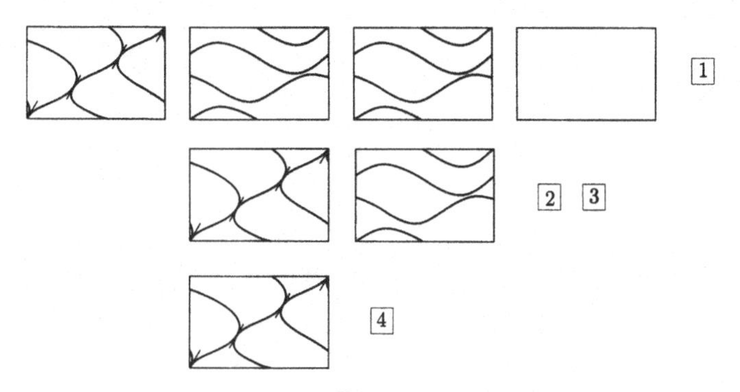

Figure 21

The proper Toda lattice. One can deduce from this proposition the topology of our original level set $\mathcal{T}_h$ (the one that was compact): the eigenvector mapping $\varphi : \mathcal{T}_h \to \text{Pic}^4(X)$ is a four-fold covering of its image, which is included in the real part of the Jacobian minus the divisors $\mathcal{D}_i$ (as φ is not injective, there might be, and in fact there are, real points in the Jacobian that are images of non-real points of $\mathcal{T}_h$). As the Hamiltonian was proper, the only values of h that can give non-empty levels $\mathcal{T}_h$ are those for which $\mathcal{T}_h'$ has a compact component (the case where $X_{\mathbf{R}}$ has the maximum possible number of components). Moreover, the signs of a_1, a_2 and a_3 (with $a_1a_2a_3 = 1$) allow us to distinguish between the four groups of connected components in $\mathcal{T}_h$. We get eventually:

2.3.4 COROLLARY. *The real level set $\mathcal{T}_h$ consists of four tori when h is such that $X_{\mathbf{R}}$ has three components, and is empty otherwise.* □

Remark. Figure 21 as well as Proposition 2.3.1 show that there is a preferred direction tangent to $\text{Jac}(X)$. Starting from an isospectral set $\mathcal{T}_h$ of Jacobi matrices (which actually has nothing to do with any differential system), we derive, using the eigenvector mapping φ, a preferred direction tangent to $\mathcal{T}_h$. This is already surprising. What is even more surprising is that this is the direction of the Toda flow, the flow of the Hamiltonian H (this can be seen in the proof of Proposition 1.2.2). See [12] for this remark.

Bifurcations. Notice that the torus can disappear and so can the cylinders, simply becoming thinner and thinner. There is of course no Morse theoretical model allowing a disc to disappear ... so they stay there.

Appendices

Appendix 1. A Poisson structure on the dual of a Lie algebra

In this appendix, I will recall rather quickly the basic definitions and constructions related to the Kirillov-Kostant structure on the dual of a Lie algebra. For details and further information, I send the reader back to the classical references on Lie groups, the beautiful book of Kirillov [48], and also e.g. the books on symplectic geometry that I have already cited, namely Libermann & Marle [59] and Audin [10]. I will use the notions (mainly the definition of a Poisson manifold) already described in the Introduction.

1.1. The Lie algebra and its dual

Remember that a (real or complex) finite-dimensional vector space $\mathfrak{g}$ is a *Lie algebra* if it is endowed with a bilinear skew-symmetric bracket

$$[\ ,\ ] : \mathfrak{g} \times \mathfrak{g} \longrightarrow \mathfrak{g}$$

that satisfies the Jacobi identity

$$[X,[Y,Z]] + [Y,[Z,X]] + [Z,[X,Y]] = 0 \qquad \forall X,Y,Z \in \mathfrak{g}.$$

Consider now the dual vector space $\mathfrak{g}^\star$. This might seem to be only a vector space (notice that the Lie bracket on $\mathfrak{g}$ defines no Lie algebra structure on $\mathfrak{g}^\star$). What the additional structure on $\mathfrak{g}$ does, though, is to define a Poisson structure (discovered and investigated mainly by Kirillov [48], Kostant [53] and Souriau [81]). If $f \in C^\infty(\mathfrak{g}^\star)$ and $\xi \in \mathfrak{g}^\star$ (I will try to use capital Latin characters for the elements of $\mathfrak{g}$ and Greek characters for the elements of $\mathfrak{g}^\star$), $df(\xi)$ is a linear form on $T_\xi\mathfrak{g}^\star = \mathfrak{g}^\star$, and we may consider it as an element of $\mathfrak{g}$ using biduality. For $f, g \in C^\infty(\mathfrak{g}^\star)$, put

$$\{f,g\}(\xi) = \langle \xi, [df(\xi), dg(\xi)] \rangle,$$

the bracket $\langle\ ,\ \rangle$ denoting the duality pairing. This defines a Poisson bracket: $\{\ ,\ \}$ is obviously skew-symmetric, it satisfies the Jacobi identity because $[\ ,\ ]$ does (as may readily be checked), and it is a derivation in each of its entries (i.e. it satisfies the Leibniz rule) because it is defined by the differentials of the functions.

Example 1. Consider the vector product on the vector space $\mathbf{R}^3$. This is a Lie algebra structure (one should check the Jacobi identity). Now identify $\mathbf{R}^3$ with its dual (using the canonical Euclidean structure) so that it becomes a Poisson manifold. Consider for instance the coordinate functions x, y and z. For a vector $v = {}^t(a, b, c)$ one has

$$\{x, y\}(v) = \left\langle v, \left[\begin{pmatrix} a \\ 0 \\ 0 \end{pmatrix}, \begin{pmatrix} 0 \\ b \\ 0 \end{pmatrix} \right] \right\rangle = abc.$$

Similar identities give

$$\{x, y\}(v) = \{y, z\}(v) = \{z, x\}(v) = abc.$$

Remark. Notice that if one were able to construct a Lie bracket on $\mathfrak{g}^\star$, this would define a Poisson structure on $\mathfrak{g}$. Conversely, certain Poisson structures on $\mathfrak{g}$ define Lie algebra structures on $\mathfrak{g}^\star$. This remark belongs to the beautiful theory of Poisson Lie groups, which is both important and useful for integrable systems – but will not be investigated in this text (see e.g. the surveys of Reyman & Semenov-Tian-Shanski [77], Semenov-Tian-Shanski [80]).

1.2. Ad, ad, Ad*, ad* and all that

There is no reason why $\mathfrak{g}^\star$ should be a symplectic variety: think of its (odd) dimension in the previous example. However, as for any Poisson manifold, it is foliated by symplectic manifolds; here these are the *coadjoint orbits*, orbits of the "coadjoint action" of the Lie group G on $\mathfrak{g}^\star$. We must begin by a few words on this notion.

Let G be a Lie group having unit 1. Denote by $\mathfrak{g}$ the tangent space T_1G. The group acts on itself by conjugation:

$$\forall g \in G, \quad \begin{array}{ccc} G & \longrightarrow & G \\ h & \longmapsto & ghg^{-1}. \end{array}$$

Now, differentiating at 1 gives the *adjoint action*

$$\forall g \in G, \quad \mathrm{Ad}_g : \begin{array}{ccc} \mathfrak{g} & \longrightarrow & \mathfrak{g} \\ X & \longmapsto & \mathrm{Ad}_g(X), \end{array}$$

which we may perfectly well consider as a map

$$\begin{array}{ccc} G & \longrightarrow & \mathrm{End}(\mathfrak{g}) \\ g & \longmapsto & \mathrm{Ad}_g, \end{array}$$

that we now differentiate at 1

$$\begin{array}{ccc} \mathfrak{g} & \longrightarrow & \mathrm{End}(\mathfrak{g}) \\ X & \longmapsto & \mathrm{ad}_X\,. \end{array}$$

Then it is an easy exercise to check that

$$\begin{array}{ccc} \mathfrak{g} \times \mathfrak{g} & \longrightarrow & \mathfrak{g} \\ (X, Y) & \longmapsto & \mathrm{ad}_X(Y) \end{array}$$

is a Lie bracket, so that $\mathfrak{g}$ is a Lie algebra, *the*[1] Lie algebra of G; we will also, at times, use the notation $[X, Y] = \mathrm{ad}_X(Y)$.

Another viewpoint on the same object: ad_X is a vector field on $\mathfrak{g}$ (whose value at Y is $[X, Y]$), the fundamental vector field[2] of the adjoint action associated with X.

Example 1 (continued). Consider the group $G = SO(3)$ of rotations in the Euclidean space $\mathbf{R}^3$. Differentiating the relation ${}^tAA = \mathrm{Id}$ at Id gives that the tangent space is the vector space $\mathfrak{g} = \mathfrak{so}(3)$ of skew-symmetric 3×3 matrices. Using the definitions above, one sees that the group G acts on $\mathfrak{g}$ by conjugation (this is the adjoint action here) and that the Lie bracket on $\mathfrak{g}$ is the commutator $[X, Y] = XY - YX$ (this will always be the case with a group of matrices). Notice that the map

$$\varphi : (x, y, z) \mapsto \begin{pmatrix} 0 & -z & y \\ z & 0 & -x \\ -y & x & 0 \end{pmatrix}$$

is a Lie algebra isomorphism (in this framework, the Jacobi identity for the vector product becomes obvious) and that conjugation of skew-symmetric matrices by rotations amounts to the canonical action of the rotations on $\mathbf{R}^3$.

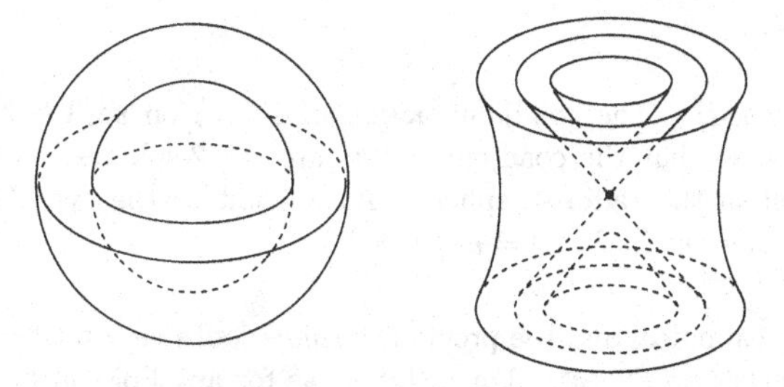

Figure 22: (Co)adjoint orbits of $SO(3)$ and $SO(2, 1)$.

Example 2. Let $G = SO(2, 1)$ be the group of isometries of the quadratic form $x^2 + y^2 - z^2$ on $\mathbf{R}^3$. Let J be the matrix of this form, so that $A \in SO(2, 1)$ if and only if ${}^tAJA = J$ and so that the Lie algebra is that of all matrices X such that ${}^tXJ + JX = 0$. It is easily checked that $\mathfrak{so}(2, 1)$ is isomorphic with $\mathbf{R}^3$ by

$$\varphi : (x, y, z) \mapsto \begin{pmatrix} 0 & -z & y \\ z & 0 & x \\ y & x & 0 \end{pmatrix}$$

and that the adjoint action is again the standard action of $SO(2, 1)$ on $\mathbf{R}^3$, so that the adjoint orbits are the sheets of the hyperboloids $x^2 + y^2 - z^2 = c$ (for $c \neq 0$), the two open cones $x^2 + y^2 - z^2 = 0$ ($z > 0$ and $z < 0$) and the point $(0, 0, 0)$. See Figure 22.

---

[1] Any Lie algebra will define (more or less uniquely, with some additional assumptions) a Lie group. When speaking of a Lie algebra $\mathfrak{g}$, I will always mean "the Lie algebra of the Lie group G" (whether the group G is explicitly mentioned or not).

[2] If the Lie group G acts on a manifold W, any vector X of the Lie algebra $\mathfrak{g}$ defines a vector field X on W, the *fundamental vector field* of the action: $X(x) = T_1 f_x(X)$, where f_x is the orbit map $g \mapsto g \cdot x$.

Let us now investigate the dual vector space $\mathfrak{g}^\star$. For $g \in G$, define

$$\mathrm{Ad}_g^\star : \mathfrak{g}^\star \longrightarrow \mathfrak{g}^\star$$

by $\langle \mathrm{Ad}_g^\star(\xi), X\rangle = \langle \xi, \mathrm{Ad}_{g^{-1}} X\rangle$ so that this is a (left) action of G, the *coadjoint action*. As above, this gives a map

$$\begin{array}{ccc} G & \longrightarrow & \mathrm{End}\,\mathfrak{g}^\star \\ g & \longmapsto & \mathrm{Ad}_g^\star \end{array}$$

that we can differentiate at 1,

$$\begin{array}{ccc} \mathfrak{g} & \longrightarrow & \mathrm{End}\,\mathfrak{g}^\star \\ X & \longmapsto & \mathrm{ad}_X^\star, \end{array}$$

so that

$$\begin{aligned} \langle \mathrm{ad}_X^\star \xi, Y\rangle &= \langle \xi, \mathrm{ad}_{-X} Y\rangle \\ &= -\langle \xi, [X,Y]\rangle. \end{aligned}$$

As above, $\mathrm{ad}_X^\star$ is a vector field on $\mathfrak{g}^\star$, the fundamental vector field of the coadjoint action associated with X.

Example 1 (continued). The coadjoint action of $SO(3)$ on $\mathfrak{so}(3)^\star \cong \mathbf{R}^3$ (see 1.5) is again by rotations, so that the coadjoint orbits are the 2-spheres centred at 0 and the only singular orbit is the radius-0 sphere. At a point v, the symplectic form of the corresponding 2-sphere is $\widetilde{\omega}_v(X,Y) = v \cdot (X \times Y)$.

All this is very canonical and the previous formula looks very much like the definition of the Poisson bracket on $\mathcal{C}^\infty(\mathfrak{g}^\star)$. For instance, as for any Poisson bracket, the one we have allows us to associate a (Hamiltonian) vector field with any function g and this is precisely $X_g(\xi) = -\,\mathrm{ad}^\star_{dg(\xi)}(\xi)$ since

$$\begin{aligned} -X_g(\xi)\cdot f = \{f,g\}(\xi) &= \langle \xi, [df(\xi), dg(\xi)]\rangle \\ &= \langle \mathrm{ad}^\star_{dg(\xi)}(\xi), df(\xi)\rangle \\ &= df(\xi)\left(\mathrm{ad}^\star_{dg(\xi)}(\xi)\right). \end{aligned}$$

1.3. Coadjoint orbits and their symplectic structures

For any $\xi \in \mathfrak{g}^\star$ define a skew-symmetric bilinear form ω_ξ on $\mathfrak{g}$ by

$$\begin{aligned} \omega_\xi(X,Y) &= \langle \xi, [X,Y]\rangle \\ &= -\langle \mathrm{ad}_X^\star \xi, Y\rangle. \end{aligned}$$

The kernel of ω_ξ is the space of all vectors X such that the vector field $\mathrm{ad}_X^\star$ vanishes at ξ.

Consider now the coadjoint orbit of ξ, and the orbit map $f_\xi : G \to \mathfrak{g}^\star$. The tangent map $T_1 f_\xi$ at 1 associates with X the value $\mathrm{ad}_X^\star(\xi)$ of the fundamental vector field of the coadjoint action. Thus ω_ξ defines a non-degenerate skew-symmetric bilinear form on the quotient vector space $\mathfrak{g}/\operatorname{Ker} T_1 f_\xi$. This is the pure linear-algebraic aspect of ω_ξ. Now comes the geometry: $T_1 f_\xi$ induces an isomorphism

$$\mathfrak{g}/\operatorname{Ker} T_1 f_\xi \longrightarrow T_\xi (G \cdot \xi)$$

so that, allowing ξ to vary in its orbit, we have defined a non-degenerate 2-form $\widetilde{\omega}$ on the orbit $\mathcal{O}$ by

$$\widetilde{\omega}_\xi (\mathrm{ad}^\star_X(\xi), \mathrm{ad}^\star_Y(\xi)) = \omega_\xi(X, Y).$$

A straightforward computation based on the Jacobi identity on $\mathfrak{g}$ and on the invariance of $\widetilde{\omega}$ shows that

$$d\widetilde{\omega} (\mathrm{ad}^\star_X, \mathrm{ad}^\star_Y, \mathrm{ad}^\star_Z) = 0 \qquad \forall X, Y, Z \in \mathfrak{g}$$

so that $d\widetilde{\omega}$ vanishes on fundamental vector fields. Since these generate the tangent space of the orbit (by definition), the 2-form $\widetilde{\omega}$ is closed, so that it defines a symplectic form on the orbit $\mathcal{O}$.

We still have to check that what we have defined is indeed the symplectic foliation of the canonical Poisson structure of $\mathfrak{g}^\star$. Consider two functions f, $g \in C^\infty(\mathfrak{g}^\star)$ and look at the Poisson bracket $\{f, g\}_{\mathcal{O}}$ of their restrictions to the given orbit $\mathcal{O}$ defined by the symplectic form on $\mathcal{O}$ that we have just described. Then:

$$\begin{aligned} df(\xi)\,(\mathrm{ad}^\star_Y(\xi)) &= \langle \mathrm{ad}^\star_Y(\xi), df(\xi)\rangle \\ &= \langle \xi, [df(\xi), Y]\rangle \\ &= \omega_\xi\,(df(\xi), Y) \\ &= \widetilde{\omega}_\xi\left(\mathrm{ad}^\star_{df(\xi)}(\xi), \mathrm{ad}^\star_Y(\xi)\right) \end{aligned}$$

so that, for the symplectic form $\widetilde{\omega}$ on the orbit $\mathcal{O}$ in which ξ lies, the Hamiltonian vector field of $f_{|\mathcal{O}}$ is

$$X^{\mathcal{O}}_{f|\mathcal{O}}(\xi) = \mathrm{ad}^\star_{df(\xi)}(\xi),$$

which is the Hamiltonian vector field of f for the canonical Poisson structure. Hence $\widetilde{\omega}$ defines a Poisson bracket on the orbit $\mathcal{O}$ such that

$$\forall f, g \in C^\infty(\mathfrak{g}^\star), \quad \left\{f_{|\mathcal{O}}, g_{|\mathcal{O}}\right\} \equiv \{f, g\}_{|\mathcal{O}},$$

and the symplectic foliation coincides with the orbit foliation.

1.4. Casimir and invariant functions

Let f be a Casimir function on $\mathfrak{g}^\star$. Let us check that it is constant on the coadjoint orbits (or invariant under the coadjoint G-action). Let X be any element of $\mathfrak{g}$ and let g_X be the linear form on $\mathfrak{g}^\star$ it defines; we then have

$$0 = \{f, g_X\}(\xi) = \langle \xi, [df(\xi), X]\rangle = \langle \mathrm{ad}^\star_X(\xi), df(\xi)\rangle$$

so that df vanishes on fundamental vector fields and thus on the tangent space of any coadjoint orbit.

Conversely, let f be an $\mathrm{Ad}^\star$-invariant function; it is constant on the orbits, thus $\{f, g_X\} = 0$ for all $X \in \mathfrak{g}$. Using Leibniz it is deduced that f commutes with all polynomials on $\mathfrak{g}^\star$ and therefore with all functions.

Thus the $\mathrm{Ad}^\star$-invariant functions on $\mathfrak{g}^\star$ coincide with the Casimir functions. In good cases (e.g. G compact or $\mathfrak{g}$ semi-simple), there are enough invariant functions to describe the generic coadjoint orbits; thus the symplectic leaves appear as the connected components of the common levels of the Casimir functions. Remember this is the case where we have been able to define the notion of integrable systems on a Poisson manifold (see 1.3 in the Introduction). The codimension of a generic orbit is the "number of invariant functions", the rank of the Lie algebra.

Remark. However, there are cases where the Casimir functions do not separate the orbits (see the counter-example below). There could even be cases where the generic leaves do not coincide with the components of the Casimir functions' levels.

Example 1 (continued). The invariant functions on $\mathbf{R}^3 = \mathfrak{so}(3)^\star$ are the functions of $x^2 + y^2 + z^2$ (one could check directly that $x^2 + y^2 + z^2$ commutes with the coordinate functions and hence with all functions). Notice that here the coadjoint orbits coincide with the levels of the Casimirs (see Figure 22).

1.5. Identification of $\mathfrak{g}^\star$ with $\mathfrak{g}$

Although there is no canonical identification of $\mathfrak{g}^\star$ with $\mathfrak{g}$, it is always possible to find an isomorphism between them, using any non-degenerate symmetric bilinear form. Here we want more; namely, we would like to identify not only the vector spaces $\mathfrak{g}$ and $\mathfrak{g}^\star$ but also the adjoint and coadjoint actions. For this, what we need is a non-degenerate symmetric bilinear form such that

$$\langle \mathrm{Ad}_g X, \mathrm{Ad}_g Y\rangle = \langle X, Y\rangle \qquad \forall g \in G, \quad \forall X, Y \in \mathfrak{g}.$$

Notice that this is equivalent to

$$\langle X, \mathrm{Ad}_{g^{-1}} Y\rangle = \langle \mathrm{Ad}_g X, Y\rangle \qquad \forall g \in G, \quad \forall X, Y \in \mathfrak{g},$$

so that the map

$$\begin{array}{ccc} \mathfrak{g} & \longrightarrow & \mathfrak{g}^\star \\ X & \longmapsto & \langle X, \cdot\rangle \end{array}$$

will interchange the Ad and $\mathrm{Ad}^\star$ actions. At the infinitesimal level, the same condition reads

$$\langle X, [Y, Z]\rangle = \langle [Z, X], Y\rangle.$$

Remark. If the Lie group G is compact, it is quite easy to manufacture an invariant non-degenerate symmetric bilinear form, starting with any positive-definite form and averaging. In general, even if there always exists an invariant symmetric form, for instance the Killing form $\langle X, Y\rangle = \mathrm{tr}(\mathrm{ad}_X \circ \mathrm{ad}_Y)$, it is not always the case that there is a non-degenerate one. However, if $\mathfrak{g}$ is *semi-simple*, the Killing form itself is non-degenerate (this is actually a possible definition of a semi-simple Lie algebra!).

Example 1 (conclusion). The canonical Euclidean structure we have used on $\mathbf{R}^3$ is invariant (!!); up to a scalar factor, this is the Killing form of $\mathfrak{so}(3)$.

Example 2 (conclusion). The quadratic form $x^2+y^2-z^2$ is of course invariant by $SO(2,1)$ (and by definition) so that the coadjoint orbits coincide with the adjoint orbits described above. Notice that the invariant functions are generated by $x^2+y^2-z^2$, so that the generic coadjoint orbits are the connected components of the common levels of the Casimirs.

Counter-example. Consider the group of triangular matrices

$$G = \left\{ \begin{pmatrix} 1 & x & z \\ 0 & 1 & y \\ 0 & 0 & 1 \end{pmatrix} \mid x, y, z \in \mathbf{R} \right\},$$

that is, the Heisenberg group $G = \mathbf{R}^3$ with multiplication

$$(x, y, z) \cdot (x', y', z') = (x + x', y + y', z + z' + xy').$$

The Lie algebra $\mathfrak{g}$ is that of upper-triangular matrices with zeros on the diagonal, which is just $\mathbf{R}^3$ (with some Lie bracket). The adjoint action (conjugation of matrices) is

$$\mathrm{Ad}_{(x,y,z)}(u, v, w) = (u, v, -yu + xv + w)$$

so that the adjoint orbits are all the lines parallel to the w-axis, except the w-axis itself, and the points of this axis. Notice that the generic orbits are one-dimensional, so they cannot be symplectic. Thus we are sure that the adjoint and coadjoint actions are *not* isomorphic, and in particular that there exists no invariant non-degenerate symmetric bilinear form on $\mathfrak{g}$.

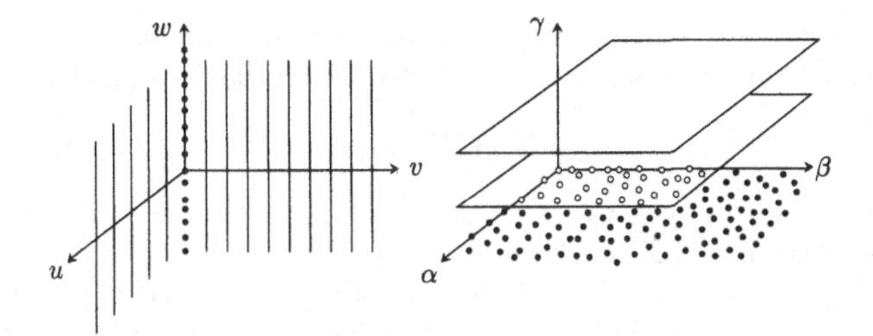

Figure 23

Let us come now to the coadjoint action:

$$\mathrm{Ad}^{\star}_{(x,y,z)}(\alpha, \beta, \gamma) = (\alpha + y\gamma, \beta - x\gamma, \gamma)$$

so that the coadjoint orbits are all the planes parallel to the $\alpha\beta$-plane except for the $\alpha\beta$-plane itself, and the points in this plane (fortunately everything is even-dimensional now). The only functions on $\mathbf{R}^3$ that can be constant on all the orbits are the functions of γ, so that the coordinate γ generates the invariants – and the Casimirs. However, the symplectic leaves are *not* the levels of γ: these are, of course, unions of orbits, but there are not enough invariants to separate the small orbits (see Figure 23).

Assume now that for some reason we have a non-degenerate symmetric bilinear form on $\mathfrak{g}$. This can also be used to define the gradient ∇f of any function f. We previously had $X_f(\xi) = \mathrm{ad}^\star_{df(\xi)}(\xi)$; here we get, similarly, $\mathrm{ad}_{\nabla_x H}\, x = [x, \nabla_x H]$ and deduce that a Hamiltonian system

$$\dot{\xi} = X_H(\xi)$$

can also be written

$$\dot{x} = [x, \nabla_x H]\,.$$

Appendix 2. R-matrices and the "AKS theorem"

Although this is a rather simple idea, originally due to Kostant [54] and Adler [1], it is quite useful. The simplest possible statement is due to Semenov-Tian-Shanski [79]. This is the one I will present here.

2.1. The R-matrix and the STS statement

Let $\mathfrak{g}$ be a Lie algebra and let R be a linear map $\mathfrak{g} \to \mathfrak{g}$. Define

$$[X,Y]_R = [RX,Y] + [X,RY]\,.$$

This is a skew-symmetric bracket. In order that it defines a Lie bracket, it has to satisfy the Jacobi identity. This is equivalent to requiring that R satisfies the equation

$$[X,[RY,RZ] - R([Y,Z]_R)] + \text{(the two analogous terms obtained by cyclic permutations of } X,Y,Z) = 0.$$

Remark. This can be considered as a variant of the Yang-Baxter equation. It is satisfied for instance when R satisfies the so-called *modified Yang-Baxter equation*:

$$[RX,RY] - R([X,Y]_R) = -[X,Y]\,.$$

See e.g. the survey of Reyman & Semenov-Tian-Shanski [77] and the lecture notes of Semenov-Tian-Shanski [80] for a general discussion.

Suppose the condition above is satisfied, so that $[\ ,\ ]_R$ is a new Lie algebra structure on $\mathfrak{g}$ and thus defines a Poisson bracket on $\mathfrak{g}^\star$, which we shall denote by $\{\ ,\ \}_R$:

$$\{\varphi,\psi\}_R(\xi) = \langle \xi, [Rd\varphi(\xi), d\psi(\xi)] + [d\varphi(\xi), Rd\psi(\xi)]\rangle.$$

The Hamiltonian vector field of φ (with respect to this Poisson structure) will be denoted by X^R_φ.

2.1.1 THEOREM (Involution theorem). *If φ is an invariant function on $\mathfrak{g}^\star$,*

$$X^R_\varphi(\xi) = \mathrm{ad}^\star_{Rd\varphi(\xi)} \cdot \xi.$$

If both φ and ψ are invariant, then

$$\{\varphi,\psi\} = \{\varphi,\psi\}_R = 0.$$

Proof. Everything is almost obvious. If φ is invariant, $\langle \xi, [d\varphi(\xi), Y] \rangle = 0$ for all $Y \in \mathfrak{g}$ so that

$$\{\varphi, \psi\}_R(\xi) = \langle \xi, [Rd\varphi(\xi), d\psi(\xi)] \rangle$$

for any function ψ. This gives the first assertion. If both functions are invariant, everything vanishes and $\{\varphi, \psi\}_R = 0$. □

Remark. We are playing with two different Lie algebra structures on $\mathfrak{g}$, but their roles are not quite symmetric. For instance, there is no reason why the Casimir functions for $\{\ ,\ \}_R$ should commute with respect to the original Poisson bracket. Also, Theorem 2.1.1 states that the differential equation

$$\dot{\xi} = \mathrm{ad}^\star_{Rd\varphi(\xi)} \cdot \xi$$

is Hamiltonian with respect to $\{\ ,\ \}_R$, but there is no reason why it should be Hamiltonian with respect to $\{\ ,\ \}$. However, just because it is Hamiltonian in the first sense, any of its solutions will stay in a common level of the Casimirs for $\{\ ,\ \}_R$. According to Theorem 2.1.1, we know that they will also stay in a coadjoint orbit of $\mathfrak{g}^\star$: the trajectories of the vector field X^R_φ lie in the intersections of the orbits in $\mathfrak{g}^\star$ and $\mathfrak{g}^\star_R$.

An interesting special case is that of a Lie algebra that is the direct sum of two Lie subalgebras:

$$\mathfrak{g} = \mathfrak{a} + \mathfrak{b}$$

(this direct sum is a direct sum of vector subspaces, *not* of Lie algebras; this means that $[a, b]$ is not necessarily zero for $a \in \mathfrak{a}$ and $b \in \mathfrak{b}$). Call P_+ and P_- the two projections, respectively onto $\mathfrak{a}$ and $\mathfrak{b}$, determined by the decomposition; write $X_\pm = P_\pm X$ and

$$R = \frac{1}{2}(P_+ - P_-).$$

In this case

$$\begin{aligned} [X, Y]_R &= \tfrac{1}{2}([X_+ - X_-, Y] + [X, Y_+ - Y_-]) \\ &= \tfrac{1}{2}([X_+ - X_-, Y_+ + Y_-] + [X_+ + X_-, Y_+ - Y_-]) \\ &= [X_+, Y_+] - [X_-, Y_-]. \end{aligned}$$

This bracket is, of course, a Lie bracket: it gives $\mathfrak{g}$ the Lie algebra structure of the direct sum $\mathfrak{a} \oplus \mathfrak{b}$ (up to sign on $\mathfrak{b}$). It is usual to denote this Lie algebra by $\mathfrak{g}_0$ and this forces us to denote $[\,,\,]_R$, $\{\,,\,\}_R$ and X^R_φ by $[\,,\,]_0$, $\{\,,\,\}_0$ and X^0_φ even if this is not very consistent. Note that if φ is an invariant function on $\mathfrak{g}^\star$,

$$0 = \mathrm{ad}^\star_{d\varphi(\xi)}(\xi) = \mathrm{ad}^\star_{d\varphi(\xi)_+}(\xi) + \mathrm{ad}^\star_{d\varphi(\xi)_-}(\xi)$$

so that

$$\mathrm{ad}^\star_{Rd\varphi(\xi)}(\xi) = -\,\mathrm{ad}^\star_{d\varphi(\xi)_-}(\xi).$$

From this, one immediately deduces:

2.1.2 COROLLARY. *Let φ be an invariant function on $\mathfrak{g}^\star$. Then*

$$X^0_\varphi(\xi) = -\,\mathrm{ad}^\star_{d\varphi(\xi)_-}(\xi).$$

If φ and ψ are two invariant functions, they commute for both Poisson brackets. □

2.2. The AKS statement

Note that Theorem 2.1.1 and Corollary 2.1.2 provide us with a systematic way of constructing Hamiltonian systems having many commuting first integrals (all the invariant functions). Next, we use them to construct Hamiltonian systems, on the dual $\mathfrak{a}^\star$ of a Lie algebra $\mathfrak{a}$ endowed with its canonical Poisson structure, that will also have many first integrals. This is the content of the Adler-Kostant-Symes theorem.

More precisely, suppose that $\mathfrak{g}$ is the Lie algebra of the group G and that A and B are subgroups of G such that the vector space $\mathfrak{g}$ is the direct sum of their Lie algebras $\mathfrak{a}$ and $\mathfrak{b}$. We consider the Lie algebra $\mathfrak{g} = \mathfrak{a} + \mathfrak{b}$ with its second Lie bracket $[\,,\,]_0$ and we try to have $\mathfrak{a}^\star$ included in $\mathfrak{g}$. For this we suppose that we are given an invariant symmetric bilinear form $\langle\,,\,\rangle$ that allows us to identify $\mathfrak{g}$ with the dual vector space $\mathfrak{g}^\star$.

Now let f be a function on $\mathfrak{g}$. Thanks to the bilinear from, it has a gradient $\nabla_x f$ ($x \in \mathfrak{g}$), which can be decomposed as $(\nabla_x f)_+ + (\nabla_x f)_-$. As ξ is the nickname of x in $\mathfrak{g}^\star$, the function f, considered as a function on $\mathfrak{g}^\star$, will be called φ, so that $d\varphi(\xi) = \nabla_x f \in \mathfrak{g}$. If f is ad-invariant, φ is ad*-invariant and

$$\begin{aligned} X^0_\varphi(\xi) &= -\operatorname{ad}^\star_{d\varphi(\xi)_-}(\xi) \quad \text{according to 2.1.2} \\ &= -\left[d\varphi(\xi)_-, x\right] \\ &= \left[x, (\nabla_x f)_-\right]. \end{aligned}$$

If we are interested in describing in this way a Hamiltonian system in $\mathfrak{a}^\star$, we must now have an inclusion $\mathfrak{a}^\star \subset \mathfrak{g}$. The isomorphism

$$\begin{array}{ccc} \mathfrak{g} & \longrightarrow & \mathfrak{g}^\star \\ x & \longmapsto & \xi \end{array}$$

sends the orthogonal subspace $\mathfrak{b}^\perp$ onto the annihilator $\mathfrak{b}^\circ$ of $\mathfrak{b}$ in $\mathfrak{g}^\star$, which in turn is identified with $\mathfrak{a}^\star$ (see Figure 24); of course, we could interchange $\mathfrak{a}$ and $\mathfrak{b}$.

Remark. As a consequence, the coadjoint orbits of A in $\mathfrak{a}^\star$ may also be seen in $\mathfrak{b}^\perp \subset \mathfrak{g} \cong \mathfrak{g}^\star$. If $g \in A \subset G$, denote by $(\operatorname{Ad}^\star g)_{\mathfrak{a}}$ its coadjoint action on $\mathfrak{a}^\star$ and by $\operatorname{Ad}^\star g$ its coadjoint action on $\mathfrak{g}^\star$ as an element of $\mathfrak{g}$. Let x be an element of $\mathfrak{b}^\perp \cong \mathfrak{a}^\star$ and $X = X_+ + X_-$ an element of $\mathfrak{g}$, so that

$$\langle (\operatorname{Ad}^\star g)_{\mathfrak{a}}(x), X_+ + X_- \rangle = \langle x, (\operatorname{Ad} g)(X_+) \rangle$$

in an obvious notation; this shows that $(\operatorname{Ad}^\star g)_{\mathfrak{a}}(x)$ is the projection on $\mathfrak{b}^\perp$ of $(\operatorname{Ad}^\star g)(x)$.

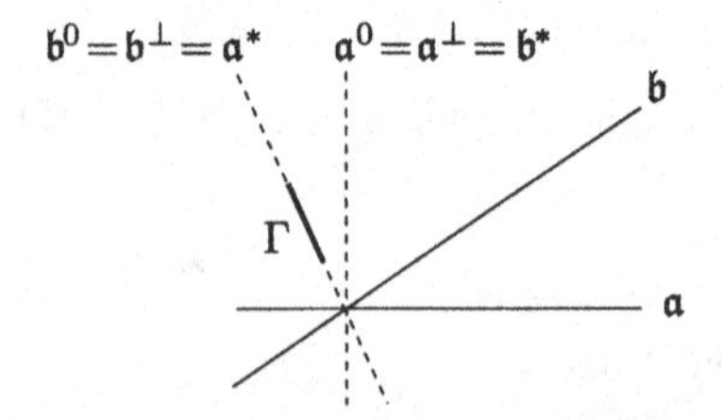

Figure 24: The AKS decomposition.

Assume now that $x \in \mathfrak{b}^\perp \cong \mathfrak{a}^\star$. Note that we do not really need f to be defined on the whole of $\mathfrak{g}$. Suppose $\Gamma \subset \mathfrak{b}^\perp$ is a union of coadjoint orbits (of $\mathfrak{a}^\star$) and suppose f to be defined on an invariant neighbourhood of Γ in $\mathfrak{g}$. As $(\nabla_x f)_- \in \mathfrak{b}$,

$$\left[x, (\nabla_x f)_-\right] \in \mathfrak{b}^\perp$$

(the invariance of the form implies that $\left[\mathfrak{b}^\perp, \mathfrak{b}\right] \subset \mathfrak{b}^\perp$), so that we stay in $\mathfrak{b}^\perp$. Moreover, $X^0_\varphi(\xi)$ is the Hamiltonian vector field for the structure $\{\ ,\ \}_0$, which is precisely the Kirillov-Kostant[3] structure on $\mathfrak{a}^\star$. Thus Corollary 2.1.2 implies:

2.2.1 COROLLARY (AKS theorem). *Let $\mathfrak{g}$ be a Lie algebra endowed with an invariant non-degenerate symmetric bilinear form. Suppose that the vector space $\mathfrak{g}$ is the direct sum of two Lie subalgebras $\mathfrak{a} + \mathfrak{b}$, so that $\mathfrak{g} = \mathfrak{a}^\perp + \mathfrak{b}^\perp$ and $\mathfrak{b}^\perp$ is identified with $\mathfrak{a}^\star$. Let $\Gamma \subset \mathfrak{b}^\perp$ be a union of coadjoint orbits in $\mathfrak{a}^\star$ and let f be an invariant function defined on a neighbourhood of Γ. The Hamiltonian system associated with f is*

$$\dot{x} = \left[x, (\nabla_x f)_-\right]$$

(the subscript minus denoting the projection onto $\mathfrak{b}$). Moreover all invariant functions defined in a neighbourhood of Γ pairwise commute. □

Remark. The geometrical situation is the following: the common levels of the invariant functions on the large Lie algebra $\mathfrak{g}$ are the (co)adjoint orbits. The flows they define on Γ preserve the intersections of Γ with the orbits, which are the common level sets of our first integrals. Note that the number of commuting first integrals we get here is bounded by the rank of the Lie algebra $\mathfrak{g}$. In order to get integrable systems, we thus have to use rather small spaces Γ.

2.3. An example

Let $\mathfrak{g} = \mathfrak{sl}_n(\mathbf{C})$ be the Lie algebra of complex traceless matrices. Let $\mathfrak{a}$ be the Lie subalgebra of lower-triangular matrices and $\mathfrak{b}$ that of skew-symmetric matrices:

$$\begin{aligned} \mathfrak{a} &= \{a \in \mathfrak{sl}_n(\mathbf{C}) \mid a_{i,j} = 0 \text{ for } j > i\} \\ \mathfrak{b} &= \{b \in \mathfrak{sl}_n(\mathbf{C}) \mid b_{j,i} = -b_{i,j}\}. \end{aligned}$$

The vector space $\mathfrak{g}$ is the direct sum of $\mathfrak{a}$ and $\mathfrak{b}$: if $X = (x_{i,j}) \in \mathfrak{sl}_n(\mathbf{C})$, then $X = X_+ + X_-$, where $X_+ = (y_{i,j}) \in \mathfrak{a}$ and $X_- = (z_{i,j}) \in \mathfrak{b}$ are defined by

$$y_{i,j} = \begin{cases} x_{i,j} + x_{j,i} & j < i \\ x_{i,j} & j = i \\ 0 & j > i \end{cases}$$

and

$$z_{i,j} = \begin{cases} -x_{i,j} & j < i \\ 0 & j = i \\ x_{i,j} & j > i. \end{cases}$$

[3]Note that the Poisson bracket of $\mathfrak{a}^\star$ may be written $\{f, g\}(x) = \langle x, [(\nabla_x f)_+, (\nabla_x g)_+]\rangle$, for $x \in \mathfrak{b}^\perp$, $(\nabla_x f)_+, (\nabla_x g)_+ \in \mathfrak{a}$.

The non-degenerate symmetric bilinear form $\langle X, Y\rangle = \operatorname{tr}(XY)$ is invariant under conjugation (this is the Killing form). It is easily checked that $\mathfrak{a}^\perp$ is the subspace of strictly lower-triangular matrices and that $\mathfrak{b}^\perp$ is that of symmetric matrices.

Consider now the invariant function $f(L) = \frac{1}{2}\operatorname{tr}(L^2)$, which is defined on $\mathfrak{g}$, and the Hamiltonian system it defines on $\mathfrak{a}^\star = \mathfrak{b}^\perp$. Since $df_L(X) = \operatorname{tr}(LX)$, $\nabla_L f = L$. For $L = (x_{i,j}) \in \mathfrak{b}^\perp$ a symmetric matrix, the projection $(\nabla_L f)_- = L_-$ onto the subspace of skew-symmetric matrices is

$$(L_-)_{i,j} = \begin{cases} -x_{i,j} & j < i \\ 0 & i = j \\ x_{j,i} & j > i. \end{cases}$$

The AKS theorem thus shows that the system $\dot{L} = [L, M]$, where L is a symmetric traceless matrix and M is the skew-symmetric matrix whose entries above the diagonal are those of L, can be understood as a Hamiltonian system and admits all the functions $L \mapsto \operatorname{tr}\left(L^k\right)$ $(2 \le k \le n)$ as pairwise-commuting first integrals. We thus have $n-1$ integrals. Let us look now for a sufficiently small space $\Gamma \subset \mathfrak{b}^\perp$.

The Lie group A is that of invertible lower-triangular matrices. Its action on $\mathfrak{b}^\perp$ (the avatar of the coadjoint action on $\mathfrak{a}^\star \cong \mathfrak{b}^\perp$) is

$$a \cdot L = \text{ projection onto } \mathfrak{b}^\perp \text{ of } aLa^{-1};$$

in other words, the symmetric matrix L is conjugated by the triangular invertible matrix a and the result aLa^{-1} is decomposed into a symmetric part $a \cdot L$ and a strictly lower-triangular part. Consider for instance the subspace $\Gamma \subset \mathfrak{b}^\perp$ of *tridiagonal* symmetric matrices (Jacobi matrices)

$$L = \begin{pmatrix} b_1 & a_1 & & & & \\ a_1 & \ddots & \ddots & & & \\ & \ddots & \ddots & \ddots & & \\ & & \ddots & \ddots & \ddots & \\ & & & \ddots & \ddots & a_{n-1} \\ & & & & a_{n-1} & b_n \end{pmatrix}$$

It is easily checked that, if $L \in \Gamma$ and $a \in A$, the entries of aLa^{-1} above the "second diagonal" are zero so that its symmetric part is still tridiagonal. Thus Γ is invariant under the A-action and we can use it in the AKS theorem: the function $f = \frac{1}{2}\operatorname{tr} L^2$ will define a Hamiltonian system on the Poisson manifold Γ of tridiagonal symmetric matrices. Precisely, this Hamiltonian system is

$$\frac{d}{dt}\begin{pmatrix} b_1 & a_1 & & \\ a_1 & \ddots & \ddots & \\ & \ddots & \ddots & a_{n-1} \\ & & a_{n-1} & b_n \end{pmatrix}$$

$$= \left[\begin{pmatrix} b_1 & a_1 & & \\ a_1 & \ddots & \ddots & \\ & \ddots & \ddots & a_{n-1} \\ & & a_{n-1} & b_n \end{pmatrix}, \begin{pmatrix} 0 & a_1 & & \\ -a_1 & \ddots & \ddots & \\ & \ddots & \ddots & a_{n-1} \\ & & -a_{n-1} & 0 \end{pmatrix}\right],$$

the celebrated (non-periodic) Toda lattice.

Note that the vector space Γ itself is *not* an orbit: the subspace of diagonal matrices is obviously closed under the A-action. However, the open subset Γ_0 of matrices in Γ with *all* non-diagonal entries non-zero (i.e. $a_i \neq 0$ for all i) *is* an orbit, that of the tridiagonal matrix

$$\begin{pmatrix} 0 & 1 & & \\ 1 & \ddots & \ddots & \\ & \ddots & \ddots & 1 \\ & & 1 & 0 \end{pmatrix}$$

(by straightforward computation). The open set Γ_0 is thus a symplectic manifold and what we have just described is a set of $n-1$ commuting functions on a $2(n-1)$ symplectic manifold (one still has to check that our integrals remain independent on Γ, but this is true).

Remark. It is possible to generalise the Toda system to other simple Lie algebras: the diagonal should be replaced by a Cartan subalgebra and the subdiagonal is related to the simple roots in $\mathfrak{sl}_n(\mathbf{C})$ (see Kostant's paper [54] and e.g. Semenov-Tian-Shanski's lecture notes [80]).

2.4. The infinite-dimensional setting

Although the previous example was both finite-dimensional and interesting, it turns out that almost all other interesting examples are infinite-dimensional, since they make essential use of Lax matrices that are *polynomials* in an additional variable λ.

Let us describe this infinite-dimensional framework. Denote by $\mathfrak{g}$ a Lie algebra of *matrices* and consider the Lie algebra $\widetilde{\mathfrak{g}}$ of Laurent polynomials in one variable λ. Because we have used biduality, the situation described above was finite-dimensional. I really do not want to enter into the details of proving that everything will actually work but this is the case in all the examples under consideration (see e.g. Adler & van Moerbeke [2] and Reyman & Semenov-Tian-Shanski [77]).

Use an invariant non-degenerate symmetric bilinear form on $\mathfrak{g}$ to define a bilinear form on $\widetilde{\mathfrak{g}}$ (with the same properties) by

$$\left\langle \sum A_i\lambda^i, \sum B_j\lambda^j \right\rangle = \operatorname{Res}\left(\sum \langle A_i, B_j\rangle \lambda^{i+j}\right)\lambda^{-k}d\lambda.$$

Suppose that a decomposition $\widetilde{\mathfrak{g}} = \mathfrak{a} + \mathfrak{b}$ is given (ignore the problems raised by duality in infinite dimensions) and an invariant function φ on $\widetilde{\mathfrak{g}}$ is given, so that we have Lax equations

$$\dot{L} = [L, M_-] \quad \text{or} \quad \dot{L} = [L, M_+]$$

where $M = \nabla_L \varphi$.

The important special case met in this book is that of the decomposition of $\widetilde{\mathfrak{g}}$ into a sum of polynomials $\sum_{i \geq 0} A_i\lambda^i$ in λ and polynomials $\sum_{i<0} A_i\lambda^i$ in λ^{-1} without constant term and of a function φ of the form

$$\varphi(A(\lambda)) = \operatorname{Res}\operatorname{tr}\left(Q(\lambda, A(\lambda))\lambda^{-1}d\lambda\right)$$

for some polynomial $Q \in \mathbf{C}[X, X^{-1}, Y]$. If this is the case, $M = d\varphi(A(\lambda)) = f(\lambda, A_\lambda)$ for $f = \partial Q/\partial Y$. Conversely, for any polynomial $f \in \mathbf{C}[X, X^{-1}, Y]$, the equation

$$\dot{L} = [L, f(\lambda, L)_+]$$

is of this form.

Remark. Note that we have much more leeway now: it is true that the AKS theorem allows us to construct Hamiltonian systems with "many" first integrals, but, in the finite-dimensional setting, there are actually *not* so very many of them, since their number is limited by the rank of the Lie algebra; this is why we were forced to look at rather small symplectic manifolds (see the example in 2.3).

Appendix 3. The eigenvector mapping and linearising flows

Following Griffiths [36], we turn back now to a very down-to-earth point of view on Lax equations (I will even try to show that there is no equation at all up to a certain point in the reasoning). Consider a Lie algebra $\mathfrak{g}$ of $N \times N$ matrices, that is, $\mathfrak{g} \subset \mathfrak{gl}(N, \mathbf{C})$.

3.1. The spectral curve and the eigenvector mapping

Consider an isospectral family of matrices $A_\lambda \in \mathfrak{g}[\lambda, \lambda^{-1}]$, in other words, the matrices in the family are Laurent polynomials in λ and all of them have the same spectrum. In particular, they all have the same characteristic polynomial.

This characteristic polynomial defines a *spectral* curve C: it might seem at first to define an affine curve

$$C_0 = \{(\lambda, \mu) \in \mathbf{C}^\star \times \mathbf{C} \mid \det(A_\lambda - \mu \operatorname{Id}) = 0\},$$

but this can be completed and normalised by adding suitable divisors over $\lambda = 0$ and ∞ (see Appendix 4), giving a curve C, which thus shows up with a covering map $\lambda : C \to \mathbf{P}^1$. Put $\mathcal{U}_+ = \lambda^{-1}(\mathbf{P}^1 - \infty)$, $\mathcal{U}_- = \lambda^{-1}(\mathbf{P}^1 - 0)$ and note that $C_0 = \mathcal{U}_+ \cap \mathcal{U}_-$. Call P_+ and P_- the divisors over 0 and ∞, so that

$$(\lambda) = P_+ - P_-.$$

Now it is possible to index the isospectral set of matrices that we are considering by the curve C, that is, to denote it by $\mathcal{T}_C$.

Suppose that the polynomials A_λ have a simple spectrum for almost all values of λ so that all the eigenspaces are lines and these lines generate the whole space $\mathbf{C}^N$: this means that $\lambda : C \to \mathbf{P}^1$ is an N-fold ramified covering map.

3.1.1 PROPOSITION. *Let A_λ be a given element of $\mathcal{T}_C$. If C is smooth, there exists a unique complex line bundle on C which is a sub-bundle of $C \times \mathbf{C}^N$ and which is such that, if $(\lambda, \mu) \in C$ and μ is a simple eigenvalue of A_λ, its fibre at the point (λ, μ) is the eigenspace of A_λ with respect to μ.*

Proof. Consider the honest algebraic subvariety V of $C \times \mathbf{P}^{N-1}(\mathbf{C})$ defined by

$$V = \{(\lambda, \mu, d) \mid d \subset \operatorname{Ker}(A_\lambda - \mu \operatorname{Id})\}$$

and the submanifold

$$V_0 = \{(\lambda, \mu, d) \in V \mid \mu \text{ is a simple eigenvalue of } A_\lambda\}.$$

Call F the set of points in C where two eigenvalues coincide. Under our assumptions, F is a finite set, the ramification locus of the finite covering map $\lambda : C \to \mathbf{P}^1$. What we need to prove is that the holomorphic one-to-one mapping

$$\begin{array}{rccc} \varphi : & C - F & \longrightarrow & V_0 \\ & (\lambda, \mu) & \longmapsto & (\lambda, \mu, d) \end{array}$$

has a holomorphic extension defined on the whole of C, with values in the closure $\overline{V_0}$ of V_0. Let A be a point in F. As C is assumed to be smooth, there exists a local chart z centred at A and, for $z \neq 0$,

$$f(z) = (z, [x_1(z), \ldots, x_N(z)])$$

in which expression the functions x_i are holomorphic except at zero where they are at worst meromorphic, so that, for m large enough (but not too large!) it is possible to extend f on a neighbourhood of 0 by

$$\varphi(z) = (z, [z^m x_1(z), \ldots, z^m x_N(z)]). \quad \square$$

Remark. I have only used that C is smooth, but it is also essential that it is a curve (the argument in the proof is strictly one-dimensional). Moreover, the vector bundle defined this way is the *unique* sub-bundle of $C \times \mathbf{C}^N$ extending the bundle of eigenvectors.

We now have the map

$$\varphi_C : \mathcal{T}_C \longrightarrow \operatorname{Pic}^d(C)$$

(for some d). It associates with any polynomial A_λ the dual L of the bundle of eigenvectors V (notice that the coordinates of the eigenvectors are sections of the dual so that L will have a positive degree d, hence this choice). We will call L the *eigenvector bundle.* The map φ_C is the *eigenvector mapping*; it can be understood (and actually was defined) as the family of mappings

$$\psi_{A_\lambda} : C \longrightarrow \mathbf{P}^{N-1}(\mathbf{C}).$$

For any given polynomial A_λ, the pull-back $\psi^\star_{A_\lambda}\mathcal{O}(1)$ of the Hopf line bundle is the eigenvector bundle.

The degree of the eigenvector bundle. The degree d is computed using the Riemann-Roch theorem. The simplest thing to do here is to consider the finite covering map $\lambda : C \to \mathbf{P}^1$. The direct image of the vector bundle is a vector bundle of rank N, whose fibre at λ is the sum of all the eigenspaces of A_λ, that is, the whole space $\mathbf{C}^N$ on which the matrix A acts (notice that this is true whether or not λ is a point in $\mathbf{P}^1$ where all eigenspaces are one-dimensional, thanks to the extension above). Now this gives d quite easily, using the Grothendieck-Riemann-Roch theorem (see A.4.2),

$$\text{(1)}\qquad \mathrm{ch}(\lambda_\star E)\,\mathrm{td}(\mathbf{P}^1) = \lambda_\star(\mathrm{ch}(E)\,\mathrm{td}(C)) \in H^\star(\mathbf{P}^1;\mathbf{Z}).$$

Since $\lambda_\star E$ is trivial of rank N, and since C and $\mathbf{P}^1$ are curves, everything else can be expressed simply in terms of the generators u of $H^2(C;\mathbf{Z})$ and t of $H^2(\mathbf{P}^1;\mathbf{Z})$. Namely, $\mathrm{ch}(E) = 1 + du$, $\mathrm{td}(C) = 1 + (1-g)u$ where g is the genus[4] of C, and $\mathrm{td}(\mathbf{P}^1) = 1 + t$ (same formula, in genus 0). Moreover, $\lambda_\star u = t$ and $\lambda_\star 1 = N$ since λ is a degree-N map, so that (1) will give

$$\begin{aligned} N(1+t) &= \lambda_\star\left[(1+du)\,(1+(1-g)u)\right] \\ &= \lambda_\star\left[1+(d-g+1)u\right] \\ &= N+(d-g+1)t \end{aligned}$$

and eventually $d = N + g - 1$.

3.2. The tangent mapping

The main result of Griffiths' paper [36] is the computation of the tangent mapping of the eigenvector mapping $\varphi_C : \mathcal{T}_C \to \mathrm{Pic}^d(C)$. Thus, what we need to understand now is the image of a tangent vector, so let us fix the latter by considering a small smooth curve $t \mapsto A_\lambda(t)$ in $\mathcal{T}_C$; we aim to compute

$$T_{A_\lambda}\varphi_C\left(\frac{d}{dt}A_\lambda(t)|_{t=0}\right),$$

that is, the infinitesimal variation of the eigenvector $L_t = \psi^\star_{A_\lambda(t)}\mathcal{O}(1)$. To save notation, let us denote $A_\lambda(0)$ by A_λ, $\psi_{A_\lambda(t)}$ by ψ_t and $\psi_{A_\lambda} = \psi_0$ by ψ.

We thus consider a local holomorphic non-vanishing section $v(x,t)$ of the vector bundle of eigenvectors (sorry, this one is $V_t = \psi_t^\star\mathcal{O}(-1) = L_t^\star$) and look at its derivative at 0.

To state the first result about this derivative, we need to introduce the exact sequence of sheaves (they are actually vector bundles) over $\mathbf{P}^{N-1}(\mathbf{C}) = P$:

$$0 \longrightarrow \mathcal{O}_P \longrightarrow \mathbf{C}^N \otimes \mathcal{O}(1) \longrightarrow TP \longrightarrow 0.$$

Recall that the inclusion map is given (in terms of vector bundles) by

$$(l,u) \longmapsto u\left((e_1^\star)|_l, \ldots, (e_N^\star)|_l\right)$$

where l is a line in $\mathbf{C}^N$ (a point of P), a vector of the fibre of $\mathcal{O}(1)$ at l is a linear form on l and $(e_1^\star, \ldots, e_N^\star)$ is the basis dual to the canonical basis of $\mathbf{C}^N$. We pull it back to C via ψ, obtaining

$$\text{(2)}\qquad 0 \longrightarrow \mathcal{O}_C \longrightarrow \mathbf{C}^N \otimes L \longrightarrow \psi^\star TP \longrightarrow 0.$$

We will use the connecting homomorphism in the associated long exact cohomology sequence

$$\longrightarrow H^0\left(C, \psi^\star TP\right) \xrightarrow{\ \delta\ } H^1\left(C;\mathcal{O}_C\right) \longrightarrow$$

recalling that $H^1(C;\mathcal{O}_C)$ is the tangent space to $\mathrm{Pic}^d(C)$ at any point (see Appendix 4). Finally we can state:

---

[4]Notice that g can be expressed in terms of N of the degrees of A_λ in λ and λ^{-1} and of the behaviour of A_λ at 0 and ∞.

3.2.1 PROPOSITION. *The derivative $\dot{v}$ at 0 of a local holomorphic non-vanishing section v of $L^\star$ defines a local holomorphic section of the N-plane bundle $\mathbf{C}^N \otimes L$ over C. Its class in the quotient $\psi^\star TP$ does not depend on the choice of v, so that it defines a global holomorphic section $[\dot{v}]$ of $\psi^\star TP$. Moreover*

$$\delta[\dot{v}] = T_{A_\lambda}\varphi_C\left(\frac{d}{dt}A_\lambda(t)|_{t=0}\right).$$

Remark. The vector space $H^0(\psi^\star TP)$ can be considered as the tangent space to the space of all deformations of the mapping $\psi : C \to P$ (the curve C is fixed). This is the reason why our actual curve $\psi_{A_\lambda}(t)$ defines an element $[\dot{v}]$ in this vector space (see the paper of Griffiths [36]).

Proof. Any section v of V defines a section α_v in the dual (simply by $\alpha_v(v) = 1$) in terms of which the inclusion $\mathbf{C} \to \mathbf{C}^N \otimes L$ is

$$(x, u) \longmapsto (x, u(v \otimes \alpha_v)).$$

Now, the derivative $\dot{v}$ at 0 is again a vector in $\mathbf{C}^N$, so that $\dot{v} \otimes \alpha_v$ is a local section of $\mathbf{C}^N \otimes L$, which actually does not depend much on v: let ρ be a non-vanishing holomorphic function and $w = \rho v$, so that $\alpha_w = \alpha_v/\rho$ and

$$\dot{w} \otimes \alpha_w = \left(\frac{\dot{\rho}}{\rho(0)}\right) v \otimes \alpha_v + \dot{v} \otimes \alpha_v.$$

The proposition follows. □

We need now to be more specific about the tangent vectors whose images we are computing.

3.2.2 PROPOSITION. *For any matrix B_λ in $\mathfrak{g}[\lambda, \lambda^{-1}]$, the matrix $[A_\lambda, B_\lambda]$ is tangent to $\mathcal{T}_C$ at A_λ.*

Remark. We assumed that C was smooth in order to be able to define the map φ_C. Notice that there is no assumption (and no claim) on the smoothness of $\mathcal{T}_C$ here. The statement is simply that $[A_\lambda, B_\lambda]$ is in the kernel of the tangent mapping to the functions defining $\mathcal{T}_C$.

Proof. The statement is equivalent to saying that the spectrum is constant on the trajectories of the vector field $A_\lambda \mapsto [A_\lambda, B_\lambda]$, in other words on the solutions of the differential equation

$$\dot{A}_\lambda = [A_\lambda, B_\lambda].$$

This is more or less obvious if one believes that solutions have the form

$$A_\lambda(t) = U(t)A_\lambda(0)U(t)^{-1}.$$

This is also an easy computation: since we want to check that $\operatorname{tr} A_\lambda^k$ is constant, consider its derivative along the flow:

$$k \operatorname{tr}\left(A_\lambda^{k-1}\dot{A}_\lambda\right) = k \operatorname{tr}\left(A_\lambda^{k-1}[A_\lambda, B_\lambda]\right) = k \operatorname{tr}[A_\lambda^k, B_\lambda] = 0. \quad \square$$

The next aim is to compute the images of these tangent vectors. I have tried to make clear, but still want to emphasise, that there has been no differential system up to now, and a fortiori no integrable system. Now, Lax equations are appearing, but recall that everything (that is, C, $\mathcal{T}_C$ and φ_C) was defined before; I do not claim, for instance, that the first integrals are in involution: there is actually no Poisson structure present here.

However, we are now considering, besides A_λ, another Laurent polynomial B_λ. Call m and n its degrees in λ^{-1} and λ respectively, so that

$$B_\lambda = \sum_{k=-m}^{n} B_k \lambda^k,$$

and call D the divisor

$$D = nP_- + mP_+.$$

Notice that D depends on B_λ (but I do not want to make the notation heavy) and is defined precisely in such a way that B_λ can be understood[5] as a homomorphism $\mathbf{C}^N \longrightarrow \mathbf{C}^N(D)$. As above, if v is a local holomorphic section of V, then $B_\lambda v$ can be viewed as a *global* section of $\mathbf{C}^N(D) \otimes L$: with the same notation as above, this is actually $(B_\lambda v) \otimes \alpha_v$, which is why one can have $B_\lambda(\rho v) = B_\lambda v$ and why I now change notation: the element of $H^0\left(C; \mathbf{C}^N(D) \otimes L\right)$ that we have just defined will be denoted by E.

Now we turn to the basic computation of Griffiths: since

$$A_\lambda(t) v_t(x) = \mu v_t(x),$$

taking derivatives with respect to t at $t = 0$ gives

$$\dot{A}_\lambda v + A_\lambda \dot{v} = \mu \dot{v}.$$

Because we are on a trajectory of $[A_\lambda, B_\lambda]$, this becomes

$$A_\lambda \left(B_\lambda v + \dot{v}\right) = \mu \left(B_\lambda v + \dot{v}\right),$$

in other words, $B_\lambda v + \dot{v}$ is an eigenvector of A_λ for the eigenvalue μ. By assumption, in general the eigenspaces of A_λ are one-dimensional, so that there exists a function g such that

$$B_\lambda v + \dot{v} = -gv.$$

Recall that we consider $B_\lambda v$ as a global section E of $\mathbf{C}^N(D) \otimes L$ and $\dot{v}$ as a global section of $\mathbf{C}^N \otimes L / \mathcal{O}_C$ and we look at a second eigenvector w such that $w = \rho v$:

$$\text{“}B_\lambda w + \dot{w}\text{”} \stackrel{\text{def}}{=} B_\lambda w \otimes \alpha_w + \dot{w} \otimes \alpha_w = B_\lambda v \otimes \alpha_v + \dot{v} \otimes \alpha_v + \frac{\dot{\rho}}{\rho} v \otimes \alpha_v$$

so that g is changed into $g + \dot{\rho}/\rho$. As ρ is a non-vanishing holomorphic function, the Laurent part of the expansion of g around D is well defined. This is what Griffiths calls the *Laurent tail* of g at D. In other words, we will consider g as a global section of the *skyscraper* sheaf $\mathcal{O}_D(D)$ defined by

$$0 \longrightarrow \mathcal{O}_C \longrightarrow \mathcal{O}_C(D) \longrightarrow \mathcal{O}_D(D) \longrightarrow 0. \qquad (3)$$

Griffiths calls this section the *residue* of B_λ and denotes it by $\rho(B)$.

[5]We should also notice that B_λ can be (and will in general be) a function of A_λ (think of a Lax equation given by the AKS theorem). When we consider the trajectories of the vector field $[A_\lambda, B_\lambda]$, B_λ will also depend on t.

Remark. Suppose that v is a local holomorphic section of V and g is a meromorphic function such that $B_\lambda v + gv$ defines a global holomorphic section of $\psi^\star TP$; then the Laurent tail of g at D is well defined and g is the residue of B_λ.

Using the connecting homomorphism ∂ of the cohomology sequence associated with (3), we get once again an element of $H^1(C; \mathcal{O}_C)$. The theorem is, of course:

3.2.3 THEOREM (Griffiths [36]). *The image of the tangent vector $[A_\lambda, B_\lambda]$ to $\mathcal{T}_C$ at A_λ under the tangent mapping $T_{A_\lambda}\varphi_C$ to the eigenvector mapping φ_C is $\partial\rho(B_\lambda) \in H^1(C; \mathcal{O}_C)$.*

Proof. Let us put together the exact sequences of sheaves (2) and (3) to get the diagram

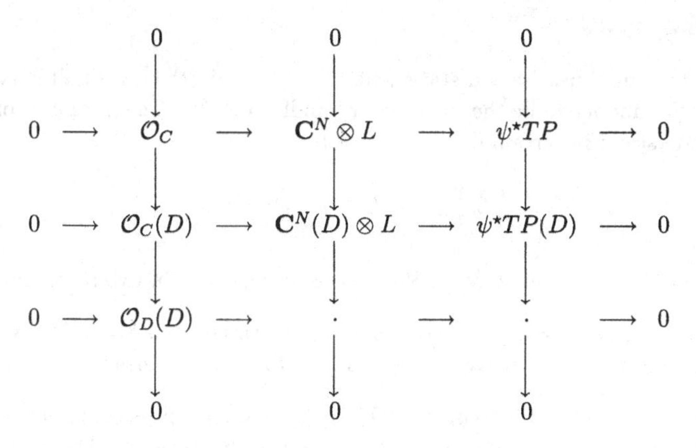

and the cohomology exact sequences

$$\begin{array}{ccccccccc}
 & & & & & & & & \downarrow \\
 & & & \longrightarrow & H^0(\mathbf{C}^N \otimes L) & \longrightarrow & H^0(\psi^\star TP) & \xrightarrow{\delta} & H^1(\mathcal{O}_C) \\
 & & & & \downarrow & & \downarrow & & \downarrow \\
\longrightarrow & H^0(\mathcal{O}_C(D)) & & \longrightarrow & H^0(\mathbf{C}^N(D) \otimes L) & \longrightarrow & \dagger & \longrightarrow & \\
 & \downarrow & & & \downarrow & & & & \\
\longrightarrow & H^0(\mathcal{O}_D(D)) & & \longrightarrow & \ddagger & & & & \\
 & \partial\downarrow & & & \downarrow & & & & \\
\xrightarrow{\delta} & H^1(\mathcal{O}_C) & & \longrightarrow & & & & & \\
 & \downarrow & & & & & & &
\end{array}$$

Because of 3.2.1, we know that the tangent vector for which we are looking is $\delta[\dot{v}] \in H^1(\mathcal{O}_C)$ and is either in the upper right-hand corner or the lower left-hand corner. With the help of B_λ, we have defined an element E in $H^0(\mathbf{C}^N(D) \otimes L)$ that has the same image as $\rho(B)$ in $\ddagger = H^0(\mathbf{C}^N \otimes L \otimes \mathcal{O}_D(D))$ and the same image as $[\dot{v}]$ in $\dagger = H^0(\psi^\star TP(D))$. By commutativity, $\delta[\dot{v}] = \partial\rho(B)$. □

Let us now turn to the special case discussed in Appendix 2 (see 2.4).

3.2.4 THEOREM (Reyman [74]). *Let $f \in \mathbf{C}[X, X^{-1}, Y]$ be a polynomial (so that $f(\lambda, \mu)$ is a holomorphic function on C_0). The image in the tangent space $H^1(C; \mathcal{O}_C)$ of the tangent vector $[A_\lambda, f(\lambda, A_\lambda)_+]$ is the class of the 1-cocycle for the covering $C = \mathcal{U}_+ \cup \mathcal{U}_-$ defined by f.*

A good exercise is to construct a simple cohomological proof analogous to that of 3.2.3. It is perhaps more illuminating to understand it, as in Reyman's original paper, as a linearisation statement involving the solutions of the Hamilton equations. This will be done in the next subsection.

3.3. Linearising flows

We turn now to some linearisation statements. Let $t \mapsto A_\lambda(t)$ be a small curve in the isospectral set $\mathcal{T}_C$, and let L_t be the eigenvector bundle of $A_\lambda(t)$. The computations above allow us to understand in certain cases the vector

$$\left(\frac{d}{dt} L_t\right)|_{t=0} \in H^1\left(C; \mathcal{O}_C\right).$$

Now, to say that φ_C linearises $A_\lambda(t)$ in $\operatorname{Pic}^d(C)$ could have one of two meanings:

- either simply that $L_t = \varphi_C\left(A_\lambda(t)\right)$ belongs to a straight line,[6] in other words that the acceleration, the derivative of the velocity dL_t/dt, is collinear to that velocity;
- or that the velocity itself is constant. This is the strongest possible requirement: not only is the curve L_t a straight line, but the parameter t itself is a linear parameter.

Notice that if one wishes to write down and/or study the behaviour of the solutions of a given differential equation, the parameter itself is meaningful. Although the above results of Griffiths apply to the discussion of both cases, I will restrict myself to the stronger property.

In the general case considered by Griffiths, we get:

3.3.1 COROLLARY. *Assume that $A_\lambda(t)$ is an integral curve of the vector field $[A_\lambda, B_\lambda]$ on $\mathcal{T}_C$. Then the necessary and sufficient condition that the image L_t in $\operatorname{Pic}^d(C)$ be linear is that*

$$\frac{d}{dt}\rho(B) \equiv 0$$

modulo the Laurent tails of the global meromorphic functions on C whose divisor of poles is $\geq D$. □

Remarks. It is possible to express the condition in terms of residues at D, and this is done very often in practice. There are applications of the theorems of 3.2 and 3.3 everywhere in this book. I suggest that sufficiently brave readers should try this on the Lax equation of III.3.3 to check that its flow is *not* linearised on the Jacobian of the spectral curve.

[6]What constitutes a straight line in a torus should be clear. If not, the next sentence in the text is a definition.

The situation in the Reyman case is the simplest one: Theorem 3.2.4 gives immediately

3.3.2 COROLLARY. *The eigenvector mapping φ_C linearises the flows of all vector fields of the form*

$$[A_\lambda, f(\lambda, A_\lambda)_+],$$

for $f \in \mathbf{C}[X, X^{-1}, Y]$.

In effect, the image of the tangent vector is the cocycle defined by f, which does not depend on the solution or, thus, on time. □

Consider the line bundle F_t defined on C by the transition function $\exp(tf)$ on C_0. An elegant way to rephrase the statements 3.2.4 and 3.3.2 is the following (actually, this is Reyman's precise statement in [74]):

3.3.3 THEOREM. *Let E_t be the eigenvector bundle of a solution $A_\lambda(t)$ of*

$$\dot{A}_\lambda = [A_\lambda, f(\lambda, A_\lambda)_+].$$

Then $E_t = E_0 \otimes F_t$.

The Lie algebra $\tilde{\mathfrak{g}} = \mathfrak{g}[\lambda, \lambda^{-1}]$ is split as a sum of Lie subalgebras $\mathfrak{a}$ of polynomials in λ and $\mathfrak{b}$ of polynomials in λ^{-1} without constant term. Let A and B be the corresponding Lie groups.

3.3.4 LEMMA. *Fix a (constant) matrix A_λ and write $M_\lambda = f(\lambda, A_\lambda)$. Suppose* $\exp(tM_\lambda)$ *is decomposed as*

$$\exp(tM_\lambda) = a(t)^{-1} b(t)$$

where a, b are paths in A and in B respectively, defined for t sufficieently small. The solution of the Hamilton equation with initial condition $A_\lambda(0) = A_\lambda$ is then

$$A_\lambda(t) = a(t) A_\lambda(0) a(t)^{-1} = b(t) A_\lambda(0) b(t)^{-1}.$$

Proof. This is just an integral version of the AKS-type theorems of Appendix 2, to which one is led after differentiation with respect to t. See, however the remark below. □

Proof of the theorem. Assuming the factorisation problem is solved, we have explicitly

$$A_\lambda(t) = a(t) A_\lambda(0) a(t)^{-1} = b(t) A_\lambda(0) b(t)^{-1}.$$

The map $a(t)$ is an automorphism of the space $\mathbf{C}^N$ on which our matrices act; it is a polynomial in $\lambda \in \mathbf{C}$. The same is true for $b(t)$, except that it is a polynomial in $\lambda^{-1} \in \mathbf{C}$. On $\mathcal{U}_+ \cap \mathcal{U}_-$, both $a(t)$ and $b(t)$ define isomorphisms between E_t and E_0. They can be compared, since

$$a(t)^{-1} b(t)|_{E_0} = \exp(tM_\lambda)|_{E_0}.$$

But E_0 is the eigenvector bundle of A_λ: for any section $v = v(\lambda, \mu)$ of E_0, $A_\lambda v = \mu v$, so that

$$M_\lambda v(\lambda, \mu) = f(\lambda, A_\lambda) v(\lambda, \mu) = f(\lambda, \mu) v(\lambda, \mu)$$

and $\exp(tM_\lambda)|_{E_0}$ is multiplication by $\exp(tf)$.

Up to now, the proof is not complete, since it relies on a solution of the factorisation problem as stated in Lemmma 3.3.4. This is known as the Riemann problem and the fact that it is not trivial can be understood from the above proof itself. Our $\exp(tM_\lambda)$ can also be considered as a gluing function for the construction of a holomorphic G-bundle of the $\mathbf{P}^1(\mathbf{C})$ in which the spectral parameter λ lives (here G is the Lie group of which the algebra of matrices $\mathfrak{g}$ is the Lie algebra). What a solution of the factorisation problem (as stated in Lemma 3.3.4) gives us is a holomorphic trivialisation of this bundle.

Now, there are non-trivial holomorphic G-bundles on $\mathbf{P}^1(\mathbf{C})$. Consider the case of the general linear group ($\mathfrak{g}$ is the Lie algebra of all $N \times N$ complex matrices). Here we can use Birkhoff's factorisation theorem, which gives

$$\exp(tM_\lambda) = a(t)^{-1} c b(t)$$

where c is a diagonal matrix of the form $(\lambda^{a_1}, \ldots, \lambda^{a_N})$, for certain integers a_i, so that the N-plane bundle constructed on $\mathbf{P}^1(\mathbf{C})$ is isomorphic to $\mathcal{O}(a_1) \oplus \cdots \oplus \mathcal{O}(a_N)$. See Chapter 8 of Pressley & Segal [72] for both a proof of Birkhoff's theorem and a discussion of the relation to bundles over $\mathbf{P}^1(\mathbf{C})$.

Clearly, here, for $t = 0$ and then for t sufficiently small, we obtain that all the a_i are zero, and this ends the proof. □

Remark. A more illuminating proof of Lemma 3.3.4 is given by Reyman [74]. Suppose we want to prove more generally that if φ is any invariant function on $\mathfrak{g}^\star$, the solutions of the Hamiltonian system associated with φ as in Theorem 2.1.1 are

$$\xi(t) = \mathrm{Ad}^\star_{b(t)}\, \xi$$

where $\xi = \xi(0)$ and $b(t) \in B$ appears in the factorisation $\exp(td\varphi(\xi)) = a(t)^{-1} b(t)$ as above. Consider $G_0 = A \times B$ and the two maps $\sigma : G_0 \to G$, $\sigma_0 : \mathfrak{g}_0 \to \mathfrak{g}$ defined by $\sigma(a, b) = ab^{-1}$, $\sigma_0(\alpha, \beta) = \alpha - \beta$. If TG and TG_0 are both trivialised by using left translations, then $d\sigma_{(a,b)} = \mathrm{Ad}_b \circ \sigma_0$ so that σ is an immersion and defines a map $\sigma^\star : T^\star G_0 \to T^\star G$ which lifts σ and which is a symplectic immersion. Now φ defines a Hamiltonian system on $T^\star G$, simply

$$\begin{cases} \dot{\xi} &= 0 \\ \dot{g} &= d\varphi(\xi), \end{cases}$$

the solutions of which are easily written:

$$\begin{cases} \xi(t) &= \xi \\ g(t) &= g \exp(td\varphi(\xi)). \end{cases}$$

The solutions of our actual system are just the projections on $\mathfrak{g}_0^\star$ of their images under $\sigma^\star$, that is, $\xi(t) = \mathrm{Ad}^\star_{b(t)}\, \xi$. □

Another remark one could make about solutions and factorisation is the following: in the finite-dimensional framework, when there is no spectral parameter λ, what we do in fact get is that the solutions can be expressed in terms of exponentials of the time variable. But we know that this is not enough for the systems of classical mechanics when, for instance, Abelian integrals appear. We have no chance of avoiding the infinite-dimensional discussion.

3.4. Miscellaneous

The regular value problem. Using the results above, it is often relatively easy to study the problem of regularity of the common levels of the first integrals (direct computation is not much fun in general). In order that the eigenvector mapping

$$\varphi_C : \mathcal{T}_C \longrightarrow \operatorname{Pic}^d(C)$$

is defined, it is enough that the spectral curve is smooth. If we are able to check that the Hamiltonian vector fields are mapped onto independent vectors of $H^1(C; \mathcal{O}_C)$, we will then be sure that they were themselves independent and *thus* that their level was regular. This argument allows us quite often to prove statements like "if C is smooth, then the corresponding level $\mathcal{T}_C$ is regular" (see [11], II.2.3.4, III.2.3.4 and IV.3.3.1); notice also that it is in general not very difficult to check whether the curve C is smooth. However, one cannot expect an absolutely general statement: there are honest examples of Lax equations obtained by honest methods in which the spectral curve C is singular for *all* values of the first integrals (see III.3.2 for instance).

The reconstruction problem. Two matrices $A \in \tilde{\mathfrak{g}} = \mathfrak{g}[\lambda, \lambda^{-1}]$ that are conjugate by an element of G (not depending on λ) have isomorphic eigenvector bundles (by an ambient isomorphism of $\mathbf{C}^N$). Under certain assumptions (which hold in particular in the Reyman case, as the higher-degree term in λ in L is diagonalisable), Reyman proves in [74] that the converse is true, in other words that it is possible to reconstruct the inverse image of a point in the image of the eigenvector mapping. He also proves that the image is the complement of a translate of the Θ-divisor. An analogous situation is described in Chapter V.

Poles of the solutions. The level sets of the invariant polynomials of $\mathfrak{g}[\lambda, \lambda^{-1}]$ are complex affine algebraic varieties. This is why one should not expect to map them *onto* complex tori (hence the missing Θ-divisor in the previous remark). From the point of view of Hamilton equations, this means that some solutions *must* go to infinity: they are meromorphic functions of the time variable. This is related to the factorisation problem mentioned in the proof of Theorem 3.3.3: once $\exp(td\varphi(\xi))$ is decomposed as $a(t)^{-1}b(t)$, the solution is holomorphic. However we explained that this was possible because t was very small. When it is no longer possible, the solution acquires a pole. Here we see that the so-called "Painlevé analysis" is intimately related to the cell decomposition of the Grassmannian. In the infinite-dimensional framework (i.e. with spectral parameter λ), this is the cell decomposition of the Sato Grassmannian, as explained by Pressley & Segal [72], but there are already illuminating examples in the finite-dimensional case (see the work of Flaschka & Haine [30] for a beautiful group-theoretical discussion of the situation in the non-periodic Toda case).

Appendix 4. Complex curves, real curves and their Jacobians

The main aim of this appendix is to describe the real part of the Jacobian of a real curve. I will explain concretely the case of genus-1 real curves. All the material here is very classical, originating in Weichold [85] and Klein [49, 50] (see also the work of Comessatti [21]; the modern reference is the paper [38] of Gross & Harris). For the convenience of the reader, I begin by recalling without proof some basic facts on (complex) algebraic curves[7] and their Jacobians used throughout this book. For further details, I send the readers back e.g. to the books of Griffiths & Harris [37], Farkas & Kra [27] and Reyssat [78].

4.1. Complex curves

A *Riemann surface* X is a compact connected complex analytic manifold of dimension 1. As such, it carries a fundamental invariant, its *genus* g, which can be given many[8] different (but equivalent) definitions. This completely describes the topology of the surface, since it is the number of handles or, more precisely, since

$$H_1(X;\mathbf{Z}) \cong \mathbf{Z}^{2g}.$$

The genus also has much to say about the analytic structure since it is the dimension of the complex vector space of holomorphic 1-forms on X, that is,

$$g = \dim H^0\left(\Omega^1_X\right),$$

where Ω^1_X is the *sheaf* (germs) of holomorphic forms and $H^0\left(\Omega^1_X\right)$ its space of (global) sections, that is, the space of holomorphic 1-forms.

An omnipresent tool is the *Riemann-Roch theorem*, which is a very precise existence theorem for meromorphic functions (see below). The equivalence between the two definitions of g above is one of its consequences.

The Riemann-Hurwitz formula. Given two Riemann surfaces X and X' and a map $f : X \to X'$, the Riemann-Hurwitz formula is an easy Euler-characteristic computation that allows us to compare the genera of X and X'. We use it quite often (without even mentioning it) in this book, mainly for degree-2 maps. In this case, it asserts that the Euler characteristic $2-2g$ of X is twice the Euler characteristic $2-2g'$ of X' minus the number of branch points.

For instance, if there is a degree-2 map $X \to \mathbf{P}^1$ branched at four points, then X has genus 1.

Completions of affine curves. Very often, especially in this book, Riemann surfaces appear as completions of affine plane curves. An *affine plane curve* C_P is the set of zeros in $\mathbf{C}^2$ of an irreducible polynomial $P \in \mathbf{C}[X,Y]$:

$$C_P = \left\{(\lambda,\mu) \in \mathbf{C}^2 \mid P(\lambda,\mu) = 0\right\}.$$

Naturally associated with C_P is a well-defined (compact) Riemann surface X_P such that there exist two finite subsets $E \subset C_P$, $F \subset X_P$ with $C_P - E = X_P - F$. The two surfaces thus coincide except at a finite number of points: X_P is compact while C_P is never so, X_P is smooth while C_P may have singularities, which are the points of E (notice that if C_P is smooth, $E = \varnothing$ and X_P is a completion of C_P).

[7] The importance of algebraic curves was well known even to Jules Verne, as the quotation opening this book shows.

[8] There are eighteen of these in Reyssat's book [78].

The Riemann surface X_P is the *normalisation*[9] of C_P. It is of course possible to complete a singular affine curve by adding *smooth points* "at infinity", that is to complete C_P without changing its singular points. This cannot be avoided when we are dealing with *families* of curves such as the ones met throughout this book, and this is what the reader can find e.g. in II.2.2, III-.2, IV.2.1, IV.3.2, V.1.2 and in Appendix 3. Very concretely, it is done simply by separating the branches of the curve at infinity, as can be seen in these examples.

Hyperelliptic curves. When a curve has a degree-2 map to $\mathbf{P}^1$, it qualifies as *hyperelliptic*. This is the case for many of the curves met in this book, that is, those that come with an equation of the form $y^2 = P(x)$ (x defines the degree-2 map). Associated with a degree-2 map is an involution, which in this case is called the hyperelliptic involution (this is $(x, y) \mapsto (x, -y)$ for $y^2 = P(x)$).

Using Riemann-Hurwitz, the genus of the curve $y^2 = P(x)$ is g if $\deg P = 2g + 1$ or $2g + 2$.

4.2. Jacobians and Picard groups

Associated with a genus-g complex curve is its Jacobian, a dimension-g complex torus, which can be described either by integration of holomorphic forms or by divisors or line bundles.

The Jacobian through holomorphic 1-forms. Let X be a complex curve. There is a natural integration mapping

$$\begin{aligned} H_1(X; \mathbf{Z}) &\longrightarrow H^0\left(\Omega_X^1\right)^* \\ \gamma &\longmapsto \left(\omega \mapsto \int_\gamma \omega\right), \end{aligned}$$

which can be shown to be injective; its image is a lattice $\Lambda \cong \mathbf{Z}^{2g}$ in $H^0\left(\Omega_X^1\right)^* \cong \mathbf{C}^g$ (still using Riemann-Roch). The *Jacobian* is the quotient torus

$$\operatorname{Jac}(X) = H^0\left(\Omega_X^1\right)^* / \Lambda.$$

The tangent space of $\operatorname{Jac}(X)$ at any point is canonically isomorphic to $H^0\left(\Omega_X^1\right)^*$, which is in turn isomorphic to $H^1(\mathcal{O}_X)$, the first cohomology group of the sheaf $\mathcal{O}_X$ of holomorphic functions on X by Serre duality (this can be described as a residue computation).

Abelian varieties. The Jacobian is not just *any* complex torus, that is, the lattice Λ is not just *any* lattice in $\mathbf{C}^g$ – it has specific features: the bilinear relations of Riemann show that the Jacobian of a curve can be embedded as an algebraic subvariety in some large projective space. This leads to the more general notion of *Abelian varieties*, complex tori that are projective varieties. I send the reader back to the books of Mumford [64] and Lange & Birkenhacke [57].

[9] This could of course be considered within the algebraic category. As this is just a summary of results, and as there are so many good books on the subject, I have tried to take the shortest way rather than the most beautiful. I will e.g. use indifferently the words "curve" and Riemann surface".

Divisors and the Picard group. Let $\operatorname{Div}(X)$ be the free Abelian group generated by the points of X. Its elements, the *divisors*, are the formal combinations

$$D = \sum_{i \in I} m_i P_i$$

(I is a finite set, $m_i \in \mathbf{Z}$, $P_i \in X$). For instance, a meromorphic function f on X defines a divisor

$$(f) = \text{ zeros of } f - \text{ poles of } f$$

(taking multiplicities into account), as does a meromorphic 1-form.

The group $\operatorname{Div}(X)$ is endowed with a natural morphism onto $\mathbf{Z}$, the *degree*:

$$\deg : \sum m_i P_i \longmapsto \sum m_i.$$

Divisors of functions have degree 0 while divisors of forms have degree $2g-2$ (another consequence of Riemann-Roch).

The group $\operatorname{Div}(X)$ is huge and not very meaningful. To add some significant information, consider the *linear equivalence* relation

$$D \sim D' \Leftrightarrow \exists \text{ a meromorphic function } f \text{ such that } D - D' = (f).$$

The quotient group is the *Picard group* $\operatorname{Pic}(X)$, its degree-d part is $\operatorname{Pic}^d(X)$ and $\operatorname{Pic}^0(X)$ is a subgroup.

The Abel-Jacobi theorem. This asserts that the integration map

$$\begin{array}{ccc} \operatorname{Div}^0(X) & \longrightarrow & H^0\left(\Omega^1_X\right)^\star \\ \Sigma(P_j - Q_j) & \longmapsto & \left(\omega \mapsto \int_{Q_j}^{P_j} \omega\right) \end{array}$$

defines an *isomorphism*

$$\operatorname{Pic}^0(X) \longrightarrow \operatorname{Jac}(X).$$

Given any point Q in X, its avatar u,

$$\begin{array}{ccc} X & \longrightarrow & \operatorname{Jac}(X) \\ P & \longmapsto & \left(\omega \mapsto \int_{Q}^{P} \omega\right), \end{array}$$

(the Abel-Jacobi map) is an injection if $g \geq 1$, and thus an isomorphism if $g = 1$: genus-1 curves are, up to the choice of a point P, groups (elliptic curves), being isomorphic to their Jacobians.

Notice that u induces an isomorphism

$$u_\star : H_1(X; \mathbf{Z}) \longrightarrow H_1(\operatorname{Jac}(X); \mathbf{Z})$$

by definition of $\operatorname{Jac}(X)$.

Divisors with nonnegative coefficients are *effective*. Notice that the effective divisors of degree d are just the points of the symmetric product

$$X^{(d)} = X^d / \mathfrak{S}_d.$$

The natural map

$$X^{(d)} \longrightarrow \operatorname{Pic}^d(X)$$

is onto for $d \geq g$. For $d = g - 1$, its image is a hypersurface in $\operatorname{Pic}^{g-1}(X)$ that is called the Θ-divisor. The various translates of this hypersurface in other degree components $\operatorname{Pic}^d(X)$ and in particular in $\operatorname{Pic}^0(X)$ or $\operatorname{Jac}(X)$ are also called Θ-divisors.

The Picard group through line bundles. There is an identification, used throughout this book, between complex line bundles and divisors. A complex line bundle on X can be defined by holomorphic (transition) functions

$$f_{\alpha,\beta} : \mathcal{U}_\alpha \cap \mathcal{U}_\beta \longrightarrow \mathbf{C}^\star$$

allowing us to glue together the trivial bundles on the open sets $\mathcal{U}_\alpha$ of a covering of X. It has global meromorphic sections (still Riemann-Roch). Such a section s can be considered as a collection of functions s_α on $\mathcal{U}_\alpha$ such that, on $\mathcal{U}_\alpha \cap \mathcal{U}_\beta$, $f_{\alpha,\beta} = s_\alpha / s_\beta$. The section s has a divisor (s) defined as the divisor of a function (zeros $-$ poles). The class of (s) in $\operatorname{Pic}(X)$, that is, modulo divisors of functions, is well-defined, depending only on the isomorphism class of the holomorphic bundle.

Notice that, in particular, the degree of (s) depends only on the bundle; this is an integer that is also its first Chern class c_1.

Conversely, any divisor (of degree d) defines a (degree-d) holomorphic line bundle, so that the Picard group can also be identified with the group of line bundles (this is actually a group, under the tensor product of line bundles). For instance, if s is a section of a line bundle L, the divisor

$$\left(\frac{1}{s}\right) = -(s)$$

is associated with the dual line bundle $L^\star$.

In slightly more pedantic terms, the cohomology exact sequence associated with the exponential

$$0 \longrightarrow \mathbf{Z} \longrightarrow \mathcal{O}_X \xrightarrow{\exp} \mathcal{O}_X^\star \longrightarrow 1$$

that is,

$$\longrightarrow H^1(X;\mathbf{Z}) \longrightarrow H^1(\mathcal{O}_X) \longrightarrow H^1(\mathcal{O}_X^\star) \xrightarrow{\deg = c_1} H^2(X;\mathbf{Z}) = \mathbf{Z},$$

gives an isomorphism

$$\operatorname{Pic}^0(X) \cong H^1(\mathcal{O}_X)/H^1(X;\mathbf{Z}).$$

The Riemann-Roch theorem. I have already used the Riemann-Roch theorem several times to state the results above. This is an evaluation of the dimension $h^0(D)$ of the complex vector space

$$\mathcal{L}(D) = \{f \mid (f) + D \geq 0\},$$

for a degree-d divisor D, namely

$$h^0(D) - h^0(K - D) = d - g + 1,$$

in which formula K, the *canonical divisor*, is the divisor of a holomorphic 1-form on X.

A divisor D such that $h^0(K-D)=0$ is *general*.

When dealing with a map between two Riemann surfaces, as we do quite often in this book, the Grothendieck-Riemann-Roch theorem is very convenient. This is not the place to discuss general definitions and/or proofs for this equality (see e.g. Hirzebruch's classical book [42]). Consider a map $f: X \to X'$ between two Riemann surfaces of respective genera g and g'. Let E be a line bundle on X. The theorem asserts that

$$\operatorname{ch}(f_\star E)\operatorname{td}(X') = f_\star(\operatorname{ch}(E)\operatorname{td}(X')).$$

In the simple one-dimensional case we are considering, this is equivalent to stating that

$$1-g'+d'=1-g+d$$

where d is the degree of the line bundle E and d' the degree of the direct image $f_\star E$.

4.3. Real structures

A real structure on a complex algebraic variety X is an anti-holomorphic involution S on X: if f is a local holomorphic function on X, $f \circ S$ is anti-holomorphic.

The basic example is of course that of $\mathbf{C}^g$, the real structure being complex conjugation. The next lemma is both a remark that we will need and an easy exercise in linear algebra:

4.3.1 LEMMA. *Let V be a complex vector space and let $S: V \to V$ be a linear anti-holomorphic involution. Let $V_{\mathbf{R}}$ be the real vector space of fixed points of S. Then (V,S) is isomorphic to $(V_{\mathbf{R}} \otimes_{\mathbf{R}} \mathbf{C}, S_0)$, where here S_0 means complex conjugation.* □

The next family of examples is the algebraic varieties described by polynomial equations with real coefficients: they carry a real structure (complex conjugation of coordinates). This is the case of the level sets and of all the curves arising from Lax matrices in the examples investigated in this book (see e.g. I.3.2.1, I.1.1, II.2.2 and II.2.1).

Note that, in this definition, a real algebraic variety is a complex algebraic variety + something else (i.e. S). This is quite different from the real part, or set of real points $X_{\mathbf{R}}$, of (X,S), which is the set of fixed points of S. There exist perfectly honest real varieties that have no real points at all and can nevertheless be used to understand real questions: the spectral curve for the symmetric top used in II.2.2 is a good example.

The real algebraic varieties in which we are interested are the real curves and their Jacobians.

4.4. Jacobians of real curves

Of course, the Jacobian of a real curve is a real variety. This can be seen both from the point of view of forms and from that of divisors.

Real structure on holomorphic 1*-forms.* The real structure S also induces an involution on functions and forms. Let f be a holomorphic function defined on an open subset $U \subset X$; then $f^S = \overline{f \circ S}$ is a holomorphic function on $S(U)$. If ω is a holomorphic 1-form on X that can be written $f(z)dz$ on U, then ω^S is the 1-form on X that can be written $\overline{f \circ S(z)}dz$ on $S(U)$.

Notice that $\frac{1}{2}\left(\omega+\omega^S\right)$ is invariant by S. Such forms are said to be *real*. If a is an integral homology class, notice also that

$$\int_{S(a)} \omega^S = \overline{\int_a \omega}. \tag{4}$$

Let (X,S) be a real curve. We have just seen S acting on the complex vector space $H^0(\Omega^1_X)$, the action being obviously anti-holomorphic. The dual vector space inherits a real structure, still called S, by

$$S\cdot\varphi = \overline{\varphi\circ S} = {}^t\overline{S}(\varphi).$$

Now, for $a\in H_1(X,\mathbf{Z})$, let φ_a be the linear form "integration on a". One has

$$S\cdot\varphi_a(\omega) = \overline{\varphi_a\circ S(\omega)} = \overline{\int_a \omega^S} = \int_{S_\star(a)} \omega,$$

using (4), so that $S\cdot F_a = F_{S_\star(a)}$ and the real structure on $H^0(\Omega^1_X)^\star$ preserves the period lattice on which it acts by the homology version $S_\star$ of S.

Real structure on divisors. There is an obvious extension of S to $\operatorname{Div}(X)$ by

$$S(\sum a_i P_i) = \sum a_i S(P_i).$$

To define a real structure on $\operatorname{Pic}(X)$, we just need to check that S is compatible with linear equivalence. But $D\sim D'$ means that there is a meromorphic function f such that $(f)=D-D'$. In this case, $S(D)-S(D')$ is the divisor of $\overline{f\circ S}$, which is a meromorphic function.

Let us denote again by S the real structure defined this way on $\operatorname{Pic}(X)$. It is quite clear that the isomorphism

$$\begin{array}{ccc} \operatorname{Pic}^0(X) & \longrightarrow & \operatorname{Jac}(X) \\ \Sigma(P_j-Q_j) & \longmapsto & \left(\omega\mapsto \int_{Q_j}^{P_j}\omega\right) \end{array}$$

is *real*, that is, compatible with real structures.

Let us now describe the real part of the Jacobian. The classical theory of Weichold and Klein asserts that the number of connected components of $\operatorname{Jac}(X)_{\mathbf{R}}$ depends only on that of $X_{\mathbf{R}}$, more precisely:

4.4.1 PROPOSITION. *Let X be a real genus-g curve. If $X_{\mathbf{R}}\neq\varnothing$, then*

$$\left|\pi_0\left(\operatorname{Jac}(X)_{\mathbf{R}}\right)\right| = 2^{|\pi_0(X_{\mathbf{R}})|-1}.$$

If X has no real point, then $\left|\pi_0\left(\operatorname{Jac}(X)_{\mathbf{R}}\right)\right|=1$ if g is even, 2 if g is odd.

Proof. We have explained how the complex vector space $H^0(\Omega^1_X)^\star$ is endowed with a real structure. According to Lemma 4.3.1, it is must be isomorphic to $\mathbf{C}^g$ with its natural real structure $z\mapsto\overline{z}$. The only thing to do is to describe the lattice Λ in $\mathbf{C}^g$. The next lemma is an exercise.

4.4.2 LEMMA. *Let Λ be a* **Z**-*module endowed with an involution S. There exists a* **Z**-*basis* $(\alpha_1, \dots, \alpha_q, \beta_1, \dots, \beta_\lambda, \gamma_{\lambda+1}, \dots, \gamma_{n-q})$ *of Λ in which the matrix of S is*

$$\begin{pmatrix} \mathrm{Id}_q & \mathrm{Id}_\lambda & 0 \\ 0 & -\mathrm{Id}_\lambda & 0 \\ 0 & 0 & -\mathrm{Id}_p \end{pmatrix}$$

(where $p = n - q - \lambda$). □

In the case of our period lattice, q must be g, as Λ generates $\mathbf{C}^g$ as a complex vector space, so that $(\alpha_1, \dots, \alpha_q)$ must be a basis of $\mathbf{R}^g$. We can now consider the complex torus $\mathbf{C}^g/\Lambda$ and its real part, that is, the image of the set of points v of $\mathbf{C}^g$ such that $\overline{v} = v$ modulo Λ. If

$$v = \sum x_i \alpha_i + \sum y_j \beta_j + \sum z_k \gamma_k$$

then

$$\overline{v} - v = \sum y_j \alpha_j - 2 \sum y_j \beta_j - 2 \sum z_k \gamma_k$$

so that $\overline{v} - v \in \Lambda$ if and only if all the y_j are integers and all the z_k are half-integers. The class of an element v can be represented as

$$\sum_{i=1}^{g} x_i \alpha_i + \sum_{k=\lambda+1}^{g} n_k \frac{\gamma_k}{2}$$

with $x_i \in [0, 1[$ and $n_k \in \{0, 1\}$. Write $m = g - \lambda$. What we have proved so far is that

$$(\mathbf{C}^g/\Lambda)_{\mathbf{R}} \text{ is diffeomorphic with } (\mathbf{R}^g/\mathbf{Z}^g) \times \{0, 1\}^m,$$

in particular that it has 2^m connected components. We must now relate m or λ (which come from the action of S on the homology of X) to the real part of X.

This is the topological part of the proof: one looks at the quotient surface $Y = X_{\mathbf{C}}/S$. This is a topological surface with boundary $X_{\mathbf{R}}$, orientable or not, and its Euler characteristic is $1 - g$. Look now at the relative homology of Y and $X_{\mathbf{R}}$ (assuming $X_{\mathbf{R}} \neq \varnothing$). From the topological viewpoint in which we are interested, the number λ is the dimension of the image of

$$\mathrm{Id} + S : H_1(X; \mathbf{Z}/2) \longrightarrow H_1(X; \mathbf{Z}/2).$$

This is the same as the image of the natural homomorphism

$$H_1(Y; \mathbf{Z}/2) \longrightarrow H_1(Y, X_{\mathbf{R}}; \mathbf{Z}/2)$$

or the kernel of the boundary homomorphism

$$\partial : H_1(Y, X_{\mathbf{R}}; \mathbf{Z}/2) \longrightarrow H_0(X_{\mathbf{R}}; \mathbf{Z}/2).$$

As ∂ is onto and $\dim H_1(Y, X_{\mathbf{R}}; \mathbf{Z}/2) = \dim H^1(Y; \mathbf{Z}/2) = g - 1$, we eventually obtain $\lambda + \dim H_0(X_{\mathbf{R}}; \mathbf{Z}/2) = g - 1$, so that $\dim H_0(X_{\mathbf{R}}; \mathbf{Z}/2) = g - \lambda - 1$.

Similarly, if X has no real point, Y is a closed non-orientable surface and X is its orientation covering, so that the Gysin exact sequence gives the desired result. □

If $X_{\mathbf{R}} \neq \varnothing$, let $X_0, \ldots, X_m$ be its connected components. Assume $m \geq 1$ ($X_{\mathbf{R}}$ has at least two components). For each $i \geq 1$, choose a path c_i' in Y with an endpoint on X_0 and the other on X_i. Lift these paths to closed curves in X and orient them to get m cycles $c_1, \ldots, c_m$ in $H_1(X; \mathbf{Z})$ such that $S_\star c_i = -c_i$. One can then easily replace the γ_i of Lemma 4.4.2 by the c_i here (that I call γ_i from now on).

We have now a surjective group homomorphism

$$\begin{array}{rcl} \psi : \operatorname{Jac}(X)_{\mathbf{R}} & \longrightarrow & (\mathbf{Z}/2)^m \\ \sum x_i \alpha_i + \sum n_k \frac{\gamma_k}{2} & \longmapsto & (n_{\lambda+1}, \ldots, n_g) \bmod 2 \end{array}$$

with a connected kernel, and such that, for any P, $Q \in X_{\mathbf{R}}$,

$$\int_Q^P \omega = -\frac{\gamma_j}{2} + \frac{\gamma_i}{2} + \text{ a real term}$$

if $P \in X_i$, $Q \in X_j$ and using the convention that $\gamma_0 = 0$.

Real part of the Picard group. There is a beautiful description of the decomposition of $\operatorname{Pic}^0(X)_{\mathbf{R}}$ as a union of its connected components, using that of $X_{\mathbf{R}}$:

4.4.3 PROPOSITION. *Assume $X_{\mathbf{R}} \neq \varnothing$ and call $X_0, \ldots, X_m$ its connected components. The map*

$$\begin{array}{rcl} X_{\mathbf{R}} & \longrightarrow & (\mathbf{Z}/2)^m \\ P & \longmapsto & \begin{cases} 0 \text{ if } P \in X_0 \\ (0, \ldots, 1_i, \ldots, 0) \text{ if } P \in X_i \end{cases} \end{array}$$

induces a surjective group homomorphism $\operatorname{Pic}^0(X)_{\mathbf{R}} \longrightarrow (\mathbf{Z}/2)^m$ *with connected kernel.*

The notation means, of course, that there is a 1 at the ith place. The proposition tells us that the connected components of $\operatorname{Pic}^0(X)_{\mathbf{R}}$ are indexed by the elements of $(\mathbf{Z}/2)^m$. Consider for instance the case of a real curve with two real components ($m = 1$). The proposition asserts that there are two possibilities for the class of a divisor $P - Q$ (P and Q are real points): either P and Q lie on the same component of $X_{\mathbf{R}}$ or not.

Proof. What we have defined is a map $X_{\mathbf{R}} \to (\mathbf{Z}/2)^m$. It defines a group homomorphism

$$\operatorname{Div}^0(X)_{\mathbf{R}} \longrightarrow (\mathbf{Z}/2)^m,$$

which is obviously onto. Notice, however, that $\operatorname{Div}^0(X)_{\mathbf{R}}$ is the group of degree-0 divisors consisting of individually real points and that we do not know a priori whether the natural map $\operatorname{Div}^0(X)_{\mathbf{R}} \to \operatorname{Pic}^0(X)_{\mathbf{R}}$ is onto.

The composition $\operatorname{Div}^0(X) \to \operatorname{Pic}^0(X) \to \operatorname{Jac}(X)$ is the integration morphism. Assume that P, Q are two real points of $X_{\mathbf{R}}$, belonging to X_i, X_j respectively. Let ω be a real 1-form. Then

$$\int_Q^P \omega = -\frac{\gamma_j}{2} + \frac{\gamma_i}{2} + \text{ a real term}$$

as we have already noticed, so that

$$\psi\left(\int_Q^P\right) = (0, \ldots, 1_i, 0 \ldots, 1_j, 0, \ldots, 0) = \varphi(P - Q).$$

Hence our diagram does commute and we are done. □

Remark. Note that this implies in particular that, for $X_{\mathbf{R}} \neq \varnothing$, $\mathrm{Div}^0(X)_{\mathbf{R}} \to \mathrm{Pic}^0(X)_{\mathbf{R}}$ is onto. This is of course not the case if X has no real point (there is an example of this situation in II.2.4).

4.5. Real genus-1 curves

In the genus-1 case, there are, according to the previous proofs, two distinct possibilities for the Jacobian of X: the period lattice is generated by 1 and

- either an element β such that $\overline{\beta} = 1 - \beta$, that is, $\beta = \frac{1}{2} + ib$, $b \in \mathbf{R}$; in this case the real part of $\mathbf{C}/\Lambda$, that is the image of the real axis, is connected
- or an element γ such that $\overline{\gamma} = -\gamma$, that is, $\gamma = ic$, $c \in \mathbf{R}$; in this case the real part of $\mathbf{C}/\Lambda$ has two connected components, the image of the real axis and that of the parallel real line through $ic/2$ (see Figure 25).

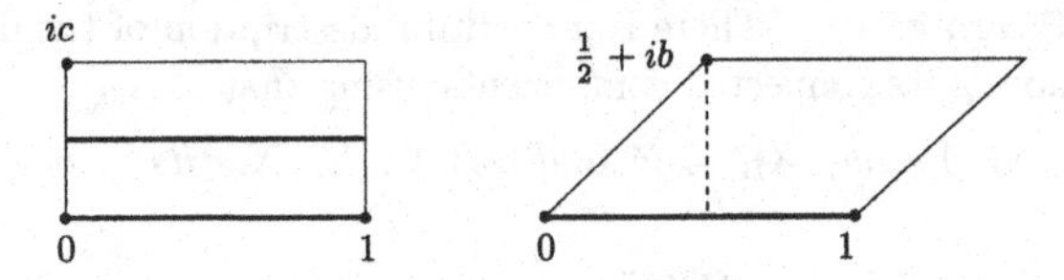

Figure 25: Real elliptic curves.

Notice that the Jacobian of X is not only a genus-1 real curve: as a Jacobian, it is a group and thus its real part is non-empty. In particular, there is no real isomorphism between a real curve without real points and its Jacobian. This is no surprise: a complex isomorphism involves a point in X. We must thus be careful in making an identification between X and $\mathbf{C}/\Lambda$.

The Abel-Jacobi mapping

$$\begin{aligned} u_{x_0} : X &\longrightarrow \mathbf{C}/\Lambda \\ x &\longmapsto \int_{x_0}^{x} \omega \end{aligned}$$

is a complex isomorphism for any point x_0 of X. Fix such a point and put

$$a = \int_{x_0}^{S(x_0)} \omega = u_{x_0}(S(x_0)).$$

Then $\overline{a} = -a \bmod \Lambda$ and, for any $y \in X$,

$$\int_{x_0}^{S(y)} \omega = \int_{x_0}^{S(x_0)} \omega + \int_{S(x_0)}^{S(y)} \omega = \overline{\int_{x_0}^{y} \omega} + a$$

so that $u_{x_0}(S(y)) = \overline{u_{x_0}(y)} + a$ for all y in X. At the $\mathbf{C}/\Lambda$ level, $z \mapsto \overline{z} + a$ is an anti-holomorphic involution, and this is the real structure of X once X and $\mathbf{C}/\Lambda$ are identified by the Abel-Jacobi mapping.

If $X_{\mathbf{R}} \neq \varnothing$, it is possible to pick an x_0 in $X_{\mathbf{R}}$; also $a = 0$ and u_{x_0} preserves real structures.

If $X_{\mathbf{R}} = \varnothing$, there is no real isomorphism. Fix an x_0 and the corresponding a and look at the involution $z \mapsto \overline{z} + a$ on $\mathbf{C}/\Lambda$. Firstly, $\overline{a} \sim -a \bmod \Lambda$ so that $a = \frac{1}{2}n + ix$ ($n \in \mathbf{Z}$, $x \in \mathbf{R}$). The case of even n is not very exciting: one can then assume that $n = 0$, the involution $z \mapsto \overline{z} + ix$ always has fixed points, so that the $X_{\mathbf{R}}$ from which we started was not really empty. One can thus assume that $n = 1$ and $a = \frac{1}{2} + ix$. Again, if $\Lambda = \langle 1, \frac{1}{2} + ib \rangle$, the involution would have fixed points. The case $X_{\mathbf{R}} = \varnothing$ thus corresponds to a lattice $\Lambda = \langle 1, ic \rangle$. To summarise:

4.5.1 PROPOSITION. *Let X be a genus-1 real curve. If $X_{\mathbf{R}} \neq \varnothing$, then X is isomorphic, as a real variety, to its Jacobian. The latter is a torus $\mathbf{C}/\Lambda$ endowed with complex conjugation, and with $\Lambda = \langle 1, ic \rangle$ ($c \in \mathbf{R}$) if $X_{\mathbf{R}}$ has two connected components and $\Lambda = \langle 1, \frac{1}{2} + ib \rangle$ ($b \in \mathbf{R}$) if $X_{\mathbf{R}}$ is connected.*

If $X_{\mathbf{R}}$ is empty, then the Jacobian of X is a torus $\mathbf{C}/\Lambda$ endowed with complex conjugation and $\Lambda = \langle 1, ic \rangle$. In particular, $\operatorname{Jac}(X)_{\mathbf{R}}$ has two connected components. Moreover, X can be identified, as a real variety, with $\mathbf{C}/\Lambda$ endowed with the involution $z \mapsto \overline{z} + \frac{1}{2} + ix$ ($x \in \mathbf{R}$). □

4.5.2 COROLLARY. *Let X be a real genus-1 curve without real points. Then, for any n, $\operatorname{Pic}^n(X)$ is a real genus-1 curve. It is isomorphic with X as a real variety (and thus has no real point) when n is odd; it is isomorphic to the Jacobian of X (as a real variety) and thus has two components when n is even.*

Proof. If P is any point of X, $P + S(P)$ is a real degree-2 divisor that can be used to identify, by translation, $\operatorname{Pic}^{2n+1}(X)$ with $\operatorname{Pic}^1(X)$ and X, and $\operatorname{Pic}^{2n}(X)$ with $\operatorname{Pic}^0(X)$ and $\operatorname{Jac}(X)$. □

The Weierstrass $\wp$-function. The Weierstrass $\wp$-function defined by the lattice Λ is:

$$\wp(z) = \frac{1}{z^2} + \sum_w{}' \left(\frac{1}{z^2 - w^2} - \frac{1}{w^2} \right).$$

It parametrises the elliptic curve $\mathbf{C}/\Lambda$: it satisfies the differential equation

$$\wp'(z)^2 = 4\wp(z)^3 - g_2\wp(z) - g_3$$

so that $\wp$, $\wp'$ define a projective embedding of $\mathbf{C}/\Lambda$ whose image is the curve

$$y^2 = 4x^3 - g_2 x - g_3.$$

Let us assume that $\mathbf{C}/\Lambda$ is real, that is, that the lattice Λ is as described above. Since it is preserved by complex conjugation,

$$\begin{aligned} \wp(\overline{z}) &= \frac{1}{\overline{z}^2} + \sum{}' \left(\frac{1}{\overline{z}^2 - w^2} - \frac{1}{w^2} \right) \\ &= \frac{1}{\overline{z}^2} + \sum{}' \left(\frac{1}{\overline{z}^2 - \overline{w}^2} - \frac{1}{\overline{w}^2} \right) \\ &= \overline{\wp(z)} \end{aligned}$$

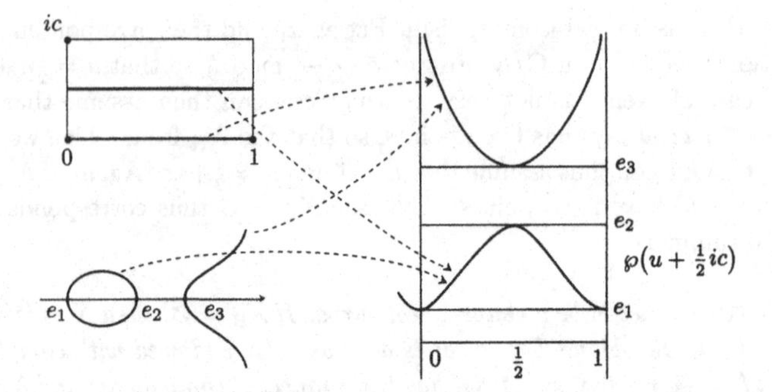

Figure 26: The Weierstrass ℘-function.

so that $\wp(z)$ is real when z is real. If the real part of $\mathbf{C}/\Lambda$ is disconnected, the points of the other component are the points $u+\frac{1}{2}ic$ ($u \in \mathbf{R}$), and they also give real values to $\wp$ as can be checked by the simple computation of $\overline{\wp\left(u+\frac{1}{2}ic\right)}$. Hence $\wp$ parametrises the real part of the curve as well. Using the classical properties of $\wp$ (see e.g. the not less classical book of Jordan [46]), that is, parity, double poles at the lattice points, periodicity and so on, we obtain the graphs in Figure 26.

Appendix 5. Prym varieties

5.1. Definition of Prym varieties

Consider a curve X endowed with an involution τ and the covering map $\pi : X \to Y = X/\tau$. It induces a homomorphism

$$\pi^\star : \operatorname{Pic}(Y) \longrightarrow \operatorname{Pic}(X)$$

that doubles degrees: just take the pre-image of a divisor. The image of the morphism $\pi^\star$ is contained in the fixed point set of $\tau^\star$. This is a subgroup of $\operatorname{Pic}(X)$, so that in particular, the image of the degree-0 component $\operatorname{Pic}^0(Y)$ is an Abelian subvariety of $\operatorname{Pic}^0(X)$.

According to a (not very hard to prove) theorem of Poincaré, there must then exist a "complementary" Abelian subvariety. This is the *Prym variety* of the involution and is denoted $\operatorname{Prym}(\tau)$ or $\operatorname{Prym}(X|Y)$. It can be defined either as the neutral component of the group of "anti-fixed" points of $\tau^\star$ or as the image of $1-\tau^\star$. From the dual point of view, it is also the quotient of the vector subspace dual to the space of $\tau^\star$ anti-invariant holomorphic 1-forms (see Mumford [65]).

Authors usually give most of their attention to the case where τ has no fixed point, in which case the covering map π is étale and the Prym variety is a principally polarised[10] Abelian variety. Here we are interested in a case where τ *has* fixed points.

This is the situation we have already met in Chapters III and IV, and which occurs frequently in the study of integrable systems with two degrees of freedom (in addition to the examples investigated in these notes, let us mention the Hénon-Heiles system (in Gérardy [34]) and the geodesics of quadrics [11]). It is the case of a double cover

[10]A Prym can also be principally polarised when it has no other choice, being one-dimensional. This is the case when X has genus 2 and τ does have two fixed points. We have met (very briefly) such a situation in III.3.2.

$\pi : C \to E$, where E is a genus-1 curve while C has genus 3, so that the corresponding involution of C has four fixed points. The Abelian variety $\text{Prym}(C|E)$ is a surface.

Following the work of Haine [39] on the dimension-4 free rigid body, Barth has investigated rather completely the geometry of these Abelian surfaces in [17]. Notice that the motivation for this "pure algebro-geometric" work came from integrable-systems questions. I invite the reader to look at this beautiful paper.

What Barth shows (and what I will use) is that if the Abelian surface A happens to contain a genus-3 curve, say D, then the involution $x \mapsto -x$ on A restricts to an involution on D and, further, that four of its fixed points are actually on D. The quotient of D by this involution is a genus-1 curve $\mathcal{E}$ and A is dual to $\text{Prym}(D|\mathcal{E})$. This is precisely the situation, studied by Haine, that we have met in IV.3.4. In that case, we had two genus-3 curves C and D and two genus-1 curves E and $\mathcal{E}$. We also had an eigenvector mapping from something that turned out to be an open subset of $\text{Prym}(D|\mathcal{E})$ and taking its values in $\text{Prym}(C|E)$. We have mentioned that the two Abelian surfaces in question are dual to each other. What I want to do now is to explain how Barth's theorem implies this result: the only thing we need to do is to show D as a submanifold of $\text{Prym}(C|E)$ with quotient $\mathcal{E}$.

5.2. Duality between two Pryms

Suppose that we are given four distinct points a_1, a_2, a_3, a_4 in $\mathbf{P}^1$ and a degree-2 polynomial P. Call b_1, b_2, b_3, b_4 the four roots of $P(\mu)^2 - \prod(\mu - a_i)$. Assume that the b_i are distinct (this is an assumption on P and on the a_i). Consider the double cover of $\mathbf{P}^1$ branched at the points a_i. This is a genus-1 curve $\mathcal{E}$. I will use, for simplicity, an equation $w^2 = \prod(\mu - a_i)$; this means that I will assume that $a_i \neq \infty$ for all i, which is always possible. We can still construct the curve D described by the equations

$$\begin{cases} w^2 &= \prod(\mu - a_i) \\ v^2 &= -w - P(\mu). \end{cases}$$

This is a double covering of $\mathcal{E}$ branched at the points of coordinates (μ, w) such that $w = -P(\mu)$, that is, $(\mu, w) = (b_i, -P(b_i))$.

At this stage, I cannot refrain from playing the following game: exchanging the a_i and b_i. This way, there is another genus-1 curve E ($z^2 = P(\mu)^2 - \prod(\mu - a_i)$) and a curve C with equations

$$\begin{cases} z^2 &= P(\mu)^2 - \prod(\mu - a_i) \\ y^2 &= -z - P(\mu). \end{cases}$$

This curve C is a double cover of E branched at the points (μ, z) such that $z = P(\mu)$, in other words $(\mu, z) = (a_i, -P(a_i))$. I will make the additional assumption that the polynomial P is not monic so that none of the b_i is at infinity. The curve E will then have two points over $\mu = \infty$ and C will have four. Call d_∞ the divisor at infinity on C (this is the pole divisor of the function μ; it has degree 4).

5.2.1 Proposition. *The symmetric product $C^{(2)}$ contains an embedded copy of D.*

Remark. As C and D play symmetric roles, an analogous statement with C and D exchanged is true.

Proof. We construct a map

$$\begin{array}{ccc} D & \longrightarrow & C^{(2)} \\ (\mu, v, w) & \longmapsto & (\mu, y_1, -P(\mu) - y_1) + (\mu, y_2, -P(\mu) - y_2). \end{array}$$

Here μ is the same on both sides and the two values y_1 and y_2 are the two roots of $y^2 - v\sqrt{2}y - w$: we are considering the (μ, y) that both belong to C and satisfy

$$\left[P(\mu) + y^2\right]^2 = z^2 = P(\mu)^2 - w^2,$$

so that $y^4 + 2y^2 P(\mu) + w^2 = 0$; but, since $P(\mu) = -v^2 - w$, the equation can be written

$$y^4 - 2y^2\left(v^2 + w^2\right) + w^2 = 0,$$

that is,

$$\left(y^2 - w\right)^2 - 2y^2 v^2 = 0,$$

in other words

$$\left(y^2 - v\sqrt{2}y - w\right)\left(y^2 + v\sqrt{2}y - w\right) = 0.$$

To summarise: once you know v and w, you are able to choose two among the four values of y that lie over μ.

The map is clearly injective: y_1 and y_2 determine

$$v = \frac{y_1 + y_2}{\sqrt{2}} \quad \text{and} \quad w = -y_1 y_2.$$

The only thing we have not proved so far is that D is *embedded*. This will be a consequence of what follows. □

Let us now map the symmetric product $C^{(2)}$ into $\operatorname{Pic}^2(C)$. As mentioned in Appendix 4, the image is a hypersurface W_2, the so-called *canonical* Θ*-divisor*. We get by composition a map $D \to \operatorname{Pic}^2(C)$.

5.2.2 PROPOSITION. *The image of D is contained in a subvariety of* $\operatorname{Pic}^2(C)$ *which is parallel to* $\operatorname{Prym}(C|E)$.

Proof. The Abelian surface $\operatorname{Prym}(C|E)$, defined as the neutral component of the anti-fixed points of $\tau : (\mu, y, z) \mapsto (\mu, -y, z)$, is a subvariety of $\operatorname{Pic}^0(C)$. Consider $\operatorname{Pic}^2(C)$ as a homogeneous space under the $\operatorname{Pic}^0(C)$-action. The subvariety mentioned in the statement of the proposition and which is parallel to $\operatorname{Prym}(C|E)$ is the component A_0 of

$$A = \left\{d \in \operatorname{Pic}^2(C) \mid \tau(d) = d_\infty - d\right\}$$

that contains the curve D. Consider in effect a point (μ_0, v, w) of D and its image d. This is the class of a divisor

$$\tilde{d} = (\mu_0, y_1, z_1) + (\mu_0, y_2, z_2)$$

with

$$\tau(\tilde{d}) = (\mu_0, -y_1, z_1') + (\mu_0, -y_2, z_2').$$

But the four points in question are the four points of D over $\mu_0 \in \mathbf{P}^1$. Thanks to the function $\mu - \mu_0$,

$$\tilde{d} + \tau(\tilde{d}) \sim d_\infty. \quad \square$$

The image of D lies at the intersection of W_2 and A_0. Now we have a map from the genus-3 curve D into the Abelian surface A_0. Thanks to the adjunction formula, we know that D must be embedded. Let us look now at the involution of A_0 acting on D: on $\mathrm{Prym}(C|E)$ we have $-x = \tau(x)$, so that the avatar of $x \mapsto -x$ of degree 2 that we are considering is, in restriction to A_0, nothing other than the involution τ. But, if y_1 and y_2 are the roots of $y^2 - v\sqrt{2}y - w$ it follows that $-y_1$ and $-y_2$ are those of $y^2 + v\sqrt{2}y - w$, so that $\tau(d)$ is the image of the point $(\mu, -v, w)$ of D: this is indeed the involution whose quotient is $\mathcal{E}$. Using eventually Barth's theorem, we have thus proved:

5.2.3 PROPOSITION. *The Abelian surfaces* $\mathrm{Prym}(C|E)$ *and* $\mathrm{Prym}(D|\mathcal{E})$ *are dual to each other.* □

5.3. If a genus-2 curve materialises

Consider again the case of the double cover of a genus-1 curve E by a genus-3 curve C, but now where the four branch points of $C \to E$ are two pairs of points that are exchanged by the involution of E. I will explain that it is then possible to construct a genus-2 curve whose Jacobian looks very much like $\mathrm{Prym}(C|E)$.

The curve E is a double covering of $\mathbf{P}^1$ branched at four points x_1, x_2, x_3, x_4; there are also two points y_1 and y_2 that are the images in $\mathbf{P}^1$ of the two pairs of ramification points of $C \to E$. We have thus singled out six points in $\mathbf{P}^1$, so that we cannot refrain from looking at the double covering X of $\mathbf{P}^1$ branched at these six points, a genus-2 curve.

5.3.1 PROPOSITION. *The curve C is the normalisation of the fibred product over $\mathbf{P}^1$ of E and X. In particular, this is an étale double covering of X. The Abelian surface* $\mathrm{Prym}(C|E)$ *is the quotient of* $\mathrm{Pic}^0(X)$ *by the order-2 subgroup generated by the class of the divisor $y_1 - y_2$.*

In this statement, y_i obviously denotes the unique point of X lying over $y_i \in \mathbf{P}^1$.

Remarks

1. This is a case where the fact that $\mathrm{Prym}(C|E)$ must have a type-$(1,2)$ polarisation should be obvious (even for those who do not know precisely what the latter means).

2. Conversely, given a genus-2 curve X, it is quite easy to construct curves C and E such that the Jacobian of X is isogenous to $\mathrm{Prym}(C|E)$: one writes X as a double covering of $\mathbf{P}^1$ branched at six points (recall that all genus-2 curves are hyperelliptic, see e.g. Farkas & Kra [27]) and divide this set of six points into two subsets of four and two points respectively.

Proof. Let C_0 be the fibred product over $\mathbf{P}^1$ of E and X:

$$C_0 = \{(e, x) \in E \times X \mid p(e) = q(x)\}.$$

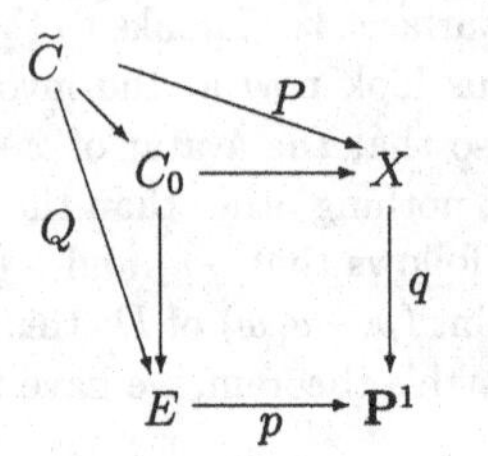

The curve C_0 is singular over the common branch points of p and q, that is, here, over the x_i. At any of these points, it actually has an ordinary double point, as can be seen from the equations, and thus it has two branches. The normalisation process is here just the separation of these two branches. The resulting normalised curve $\widetilde{C}$ is a double covering of E, branched at the points of E that lie over y_1 and $y_2 \in \mathbf{P}^1$ (over each point x_i we indeed have two points), thus $\widetilde{C}$ is C; moreover, this is also a double covering of X, but now without branching points.

Considering the two involutions on C, it is easily checked that the images of $H^0(\Omega^1_E)$ and $H^0(\Omega^1_X)$ are complementary vector subspaces in $H^0(\Omega^1_C)$: the former consists of fixed points of the involution giving the quotient E, the latter consists of those of the involution defining X. Thus the image of $\mathrm{Pic}^0(X)$ in $\mathrm{Pic}^0(C)$ is complementary to that of $\mathrm{Pic}^0(E)$. Hence, to finish the proof of the proposition it it is enough to check that this image is the quotient of $\mathrm{Pic}^0(X)$ by the right order-2 subgroup, that is, that this subgroup is the kernel of $Q^\star : \mathrm{Pic}^0(X) \to \mathrm{Pic}^0(C)$. This is a consequence of the following lemma, which we have already used in III.2.3.

5.3.2 LEMMA. *Let Z be a smooth curve and $\mathcal{D}$ be an element of order 2 in $\mathrm{Pic}^0(Z)$. Let $\pi : Y \to Z$ be the étale double cover associated with $\mathcal{D}$. The kernel of*

$$\pi^\star : \mathrm{Pic}^0(Z) \longrightarrow \mathrm{Pic}^0(Y)$$

is the subgroup generated by $\mathcal{D}$.

Let us delay the proof of the lemma. It may be assumed, and I will suppose for simplicity, that x_i, $y_j \neq \infty \in \mathbf{P}^1$. The curve X then has two points ∞_+ and ∞_- over $\infty \in \mathbf{P}^1$ and it is possible to write an (affine) equation of X in the form

$$u^2 = \prod_{i=1}^{4}(x - x_i)\prod_{j=1}^{2}(x - y_j)\,.$$

In this way x and u are two meromorphic functions on X and

$$((x - y_1)(x - y_2)) = 2(y_1 + y_2) - 2(\infty_+ + \infty_-)$$

so that the linear equivalence class $\mathcal{D}$ of $(y_1 + y - 2) - (\infty_+ + \infty_-)$ is an order-2 element of $\mathrm{Pic}^0(X)$, and this is clearly the element that defines the covering: C may be considered as the Riemann surface of $\sqrt{(x - y_1)(x - y_2)}$ over X (in other words

$$\begin{cases} u^2 &= \prod(x - x_i)\prod(x - y_j) \\ v^2 &= \prod(x - y_j) \end{cases}$$

constitute a system of equations for C_0). Now,

$$y_1 + y_2 - \infty_+ - \infty_- \sim y_1 - y_2$$

on X since $2y_2 - \infty_+ - \infty_-$ is the divisor of the meromorphic function $x - y_2$. □

Proof of the lemma. This is more or less clear using the same ideas, but I will give a topological proof. The set of order-2 elements in $\text{Pic}^0(Z)$ is naturally identified with $H^1(Z;\mathbf{Z})/2H^1(Z;\mathbf{Z})$, that is, with $H^1(Z;\mathbf{Z}/2)$ or $\text{Hom}(\pi_1(Z);\mathbf{Z}/2)$. In this way, an order-2 element $\mathcal{D}$ of $\text{Pic}^0(Z)$ defines a covering $Y \to Z$ of Riemann surfaces: if $2\mathcal{D} = (f)$, the curve Y has the equation $y^2 = f$ in $\mathbf{C}\times X$. From the topological viewpoint, this is nothing other than the covering defined by the index-2 subgroup of $\pi_1(Z)$ that is the kernel of the avatar of $\mathcal{D}$ in $\text{Hom}(\pi_1(Z);\mathbf{Z}/2)$. So we have an exact sequence

$$1 \longrightarrow \pi_1(Y) \xrightarrow{\pi_\star} \pi_1(X) \xrightarrow{D} \mathbf{Z}/2 \longrightarrow 0$$

from which we deduce the lemma by duality. □

The Kowalevski curve X. In the investigation of the Kowalevski rigid body in Chapter III, we have met several curves:

- on the one hand a genus-2 curve X and an elliptic curve $\mathcal{E}'$ (III.1); according to Kowalevski, the solutions may be expressed in terms of ϑ-functions associated with X;
- on the other hand a genus-3 curve C and an elliptic curve E (III.2); according to Bobenko, Reyman and Semenov-Tian-Shanski, the solutions may be expressed in terms of ϑ-functions associated with $\text{Prym}(C|E)$.

There must thus be some relations, more precisely correspondences, between these two groups of curves, but they seem to be not well understood. Let me make two remarks.

- The covering $C \to E$ does not fit very well into Proposition 5.3.1: the branch points in E are not exchanged by the elliptic involution (see III.2.2.1).
- The curve $\mathcal{E}'$ is nevertheless a double covering of $\mathbf{P}^1$, branched at the three roots e_1, e_2, e_3 of the Kowalevski polynomial φ and at ∞, and X has the equation

$$y^2 = -2(x-b)(x-c)\varphi(x),$$

where, to simplify notation, b, $c = H \pm \sqrt{K}$ (in the notation of III.1). Hence X is a double covering of $\mathbf{P}^1$, branched at e_1, e_2, e_3, b, c and ∞, in particular at the branch points of $\mathcal{E}' \to \mathbf{P}^1$, so that it is possible to apply Proposition 5.3.1 to X and $\mathcal{E}'$. We get a genus-3 curve $\widetilde{X}$ whose Jacobian contains both $\mathcal{E}'$ and a quotient $\text{Pic}^0(X)/\langle D\rangle = \text{Prym}(\widetilde{X}|\mathcal{E}')$.

References

[1] M. ADLER, On a trace functional for formal pseudodifferential operators and the symplectic structure for the Korteweg-de Vries type equations, *Invent. Math.* **50** (1979), 451–500.

[2] M. ADLER & P. VAN MOERBEKE, Completely integrable systems, Euclidean Lie algebras and curves *and* Linearization of Hamiltonian systems, Jacobi varieties and representation theory, *Adv. Math.* **38** (1980), 267–317 *and* 318–379.

[3] M. ADLER & P. VAN MOERBEKE, The algebraic integrability of geodesic flow on $SO(4)$, *Invent. Math.* **67** (1982), 297–331.

[4] M. ADLER & P. VAN MOERBEKE, The Kowalevski and Hénon-Heiles motions as Manakov geodesic flows on $SO(4)$—a two-dimensional family of Lax pairs, *Commun. Math. Phys.* **113** (1988), 659–700.

[5] M. ADLER & P. VAN MOERBEKE, The complex geometry of the Kowalevski-Painlevé analysis, *Invent. Math.* **97** (1989), 3–51.

[6] M. ADLER & P. VAN MOERBEKE, The Toda lattice, Dynkin diagrams, singularities and Abelian varieties, *Invent. Math.* **103** (1991), 223–278.

[7] P. APPELL, *Traité de mécanique rationnelle,* vol. II, Gauthier-Villars, Paris (1896).

[8] G. G. APPELROT, Не вполне симметричные тяжелые гироскопы, in *Volume in homage to Kowalevskaya*, Moscow (1940).

[9] V. I. ARNOLD, Математические методы классической механики, Наука, Москва, 1974, *Mathematical methods of classical mechanics*, Graduate Texts in Math., Springer (1978).

[10] M. AUDIN, *The topology of torus actions on symplectic manifolds*, Progr. Math. **93**, Birkhäuser (1991).

[11] M. AUDIN, Courbes algébriques et systèmes intégrables: géodésiques des quadriques, *Expositiones Math.* **12** (1994), 193–226.

[12] M. AUDIN, Vecteurs propres de matrices de Jacobi, *Ann. Institut Fourier* **44** (1994), 1505–1517.

[13] M. AUDIN, Topologie des systèmes de Moser en dimension 4, in *Volume in homage to Floer*, Progr. Math. **133**, Birkhäuser (1995).

[14] M. AUDIN, Lectures on integrable systems and gauge theory, *to appear* (1995).

[15] M. AUDIN & R. SILHOL, Variétés abéliennes réelles et toupie de Kowalevski, *Compositio Math.* **87** (1993), 153–229.

[16] O. BABELON, P. CARTIER & Y. KOSMANN-SCHWARZBACH, eds., *Lectures on integrable systems*, World Scientific (1994).

[17] W. BARTH, Abelian surfaces with $(1,2)$-polarization, *Adv. Stud. Pure Math.* **10** (1987), 41–84.

[18] A. I. BOBENKO, A. G. REYMAN & M. A. SEMENOV-TIAN-SHANSKY, The Kowalevski top 99 years later: A Lax pair, generalizations and explicit solutions, *Commun. Math. Phys.* **122** (1989), 321–354.

[19] A. I. BOBENKO & V. B. KUZNETSOV, Lax representation and new formulae for the Goryachev-Chaplygin top, *J. Phys. A* **21** (1988), 1999–2006.

[20] N. BOURBAKI, *Algèbre, chapitre 9*, Hermann, Paris (1959).

[21] A. COMESSATTI, Sulle varietà abeliane reale I *e* II, *Ann. Mat. Pura Appl.* **2** (1924), 67–106 *e* **4** (1926), 27–71.

[22] R. CUSHMAN & H. KNÖRRER, The momentum mapping of the Lagrange top, in *Differential geometric methods in physics,* H. Doebner et al. eds., Springer Lecture Notes in Math. **1139** (1985).

[23] R. CUSHMAN & J.C. VAN DER MEER, The Hamiltonian Hopf bifurcation in the Lagrange top, in *Géométrie symplectique et mécanique, Proceedings 1988,* C. Albert ed., Springer Lecture Notes in Math. **1416** (1990).

[24] N. DESOLNEUX-MOULIS, Dynamique des systèmes hamiltoniens complètement intégrables sur les variétés compactes, in *Géométrie symplectique et mécanique, Proceedings 1988,* C. Albert ed., Springer Lecture Notes in Math. **1416** (1990).

[25] E. DE DINTEVILLE, La toupie, *Bull. de l'Institut de Ling. de Louvain* **98** (1973), 405–413.

[26] B. DUBROVIN, Theta functions and nonlinear equations, *Russian Math. Surveys* **36** (1981), 11–92.

[27] H. FARKAS & I. KRA, *Riemann surfaces*, Graduate Texts in Math. **71**, Springer (1980).

[28] H. FLASCHKA, The Toda lattice I. Existence of integrals, *Phys. Res. B* **9** (1974), 1924–1925.

[29] H. FLASCHKA, Integrable systems and torus actions, *in* [16].

[30] H. FLASCHKA & L. HAINE, Variétés de drapeaux et réseaux de Toda, *Math. Z.* **208** (1991), 545–556.

[31] A. T. FOMENKO, The topology of surfaces of constant energy in integrable Hamiltonian systems and obstructions to integrability, *Math. USSR Izvestya* **29** (1987), 629–658.

[32] A. T. FOMENKO, ed., *Topological classification of integrable systems*, Advances in Soviet Math., **10**, Amer. Math. Soc. (1991).

[33] L. GAVRILOV & A. ZHIVKOV, The Lagrange top, *preprint* (1995).

[34] S. GÉRARDY, Le système de Hénon-Heiles, *preprint* (1995).

[35] V. V. GOLUBEV, *Lectures on integration of the equations of motion of a rigid body about a fixed point*, Israel program for scientific translations, Haifa (1960).

[36] P. A. GRIFFITHS, Linearizing flows and a cohomological interpretation of Lax equations, *Amer. J. of Math.* **107** (1985), 1445–1483.

[37] P. A. GRIFFITHS & J. HARRIS, *Principles of algebraic geometry*, Wiley (1978).

[38] B. GROSS & J. HARRIS, Real algebraic curves, *Ann. Scient. Ec. Norm. Sup.* **14** (1981), 157–182.

[39] L. HAINE, Geodesic flow on $SO(4)$ and abelian surfaces, *Math. Ann.* **263** (1983), 435–472.

[40] L. HAINE & E. HOROZOV, A Lax pair for Kowalevski's top, *Physica D* **29** (1987), 173–180.

[41] D. HILBERT & S. COHN-VOSSEN, *Anschaulige Geometrie*, Springer (1932).

[42] F. HIRZEBRUCH, *Neue topologische Methoden in der algebraischen Geometrie*, Ergebnisse der Math. und ihrer Grenzgebiete, Springer (1962); *New topological methods in algebraic geometry*, die Grundlehren der Math. Wiss. in Einzeldarstellungen, Springer (1966).

[43] E. HOROZOV & P. VAN MOERBEKE, The full geometry of Kowalevski's top and $(1,2)$-abelian surfaces, *Comm. Pure and Appl. Math.* **42** (1989), 357–407.

[44] E. HUSSON, Recherche des intégrales algébriques dans le mouvement d'un solide pesant autour d'un point fixe, *Ann. Fac. Sci. Toulouse* **8** (1906), 73–152.

[45] C. JACOBI, *Vorlesungen über Dynamik*, Gesammelte Werke, Supplementband, Berlin (1884).

[46] C. JORDAN, *Cours d'analyse de l'Ecole Polytechnique,* tome II, 3e édition, Gauthier-Villars, Paris (1959).

[47] M. P. KHARLAMOV, Bifurcation of common levels of first integrals of the Kowalevskaya problem, *PMN USSR* **47** (1983), 737–743.

[48] A. A. KIRILLOV, Элементы теории представлений, Наука, Москва, 1971, *Elements of the theory of representations*, Grundlehren der math. Wissenschaften, Springer (1976).

[49] F. KLEIN, *On Riemann's theory of algebraic functions and their integrals*, Macmillan and Bowes, Cambridge (1893).

[50] F. KLEIN, Ueber Realitätverhältnisse beider einem beliebigen Geschichte zugehoren Normalcurve der φ, *Math. Ann.* **42** (1893), 1–29.

[51] F. KLEIN & A. SOMMERFELD, *Theorie des Kreisels*, Teubner, Leipzig (1897).

[52] H. KNÖRRER, Geodesics on the ellipsoid, *Invent. Math.* **39** (1980), 119–143.

[53] B. KOSTANT, Quantization and representation theory I : prequantization, in *Lectures in modern analysis and applications III*, Lecture Notes in Math., **170**, Springer.

[54] B. KOSTANT, The solution to a generalized Toda lattice and representation theory, *Adv. in Math.* **34** (1979), 195–338.

[55] S. KOWALEVSKI, Sur le problème de la rotation d'un corps solide autour d'un point fixe, *Acta Math.* **12** (1889), 177–232.

[56] J. L. LAGRANGE, *Mécanique Analytique*, Edition de la librairie Scientifique et Technique Albert Blanchard, Paris (1965).

[57] H. LANGE & CH. BIRKENHAKE, *Complex Abelian varieties*, Grundlehren der math. Wissenschaften, Springer (1992).

[58] L. LERMAN & YA. UMANSKII, Structure of the Poisson action of $\mathbf{R}^2$ on a four dimensional symplectic manifold I *and* II, *Selecta Math. Sov.* **6** (1987), 365–396 *and* **7** (1988) 39–48.

[59] P. LIBERMANN & C.-M. MARLE, *Symplectic geometry and analytic mechanics*, Math. and its Appl. **35**, Reidel, Boston (1987).

[60] S. V. MANAKOV, Note on the integration of Euler's equations of the dynamics of an n-dimensional rigid body, *Funct. Anal. Appl.* **11** (1976), 328–329.

[61] J. MARSDEN & A. WEINSTEIN, Reduction of symplectic manifolds with symmetry, *Rep. Math. Phys.* **5** (1974), 121–130.

[62] P. VAN MOERBEKE & D. MUMFORD, The spectrum of difference operators and algebraic curves, *Acta Math.* **143** (1979), 93–154.

[63] J. MOSER, Geometry of quadrics and spectral theory, in *The Chern Symposium*, *Springer* (1980), 147–188.

[64] D. MUMFORD, *Abelian varieties*, Tata Institute, Bombay (1970).

[65] D. MUMFORD, Prym varieties, in *Contributions to analysis (a collection of papers dedicated to Lipman Bers)*, Academic Press, New York (1974), 325–350.

[66] D. MUMFORD, *Tata Lectures on Theta II*, Progr. Math. **43**, Birkhäuser (1984).

[67] NGUYEN T. D., *The symplectic topology of integrable Hamiltonian systems, thèse*, Université de Strasbourg (1994).

[68] A.A. OSHEMKOV, The topology of surfaces of constant energy and bifurcation diagrams for integrable cases of the dynamics of a rigid body on $so(4)$, *Russian Math. Surveys* **42** (1987), 241–243.

[69] A. M. PERELOMOV, Lax representations for the systems of S. Kowalewskaya type, *Commun. Math. Phys.* **81** (1981), 239–244.

[70] A. M. PERELOMOV, *Integrable systems of classical mechanics and Lie algebras*, Birkhäuser (1990).

[71] L. PIOVAN, Cyclic coverings of Abelian varieties and the Goryachev-Chaplygin top, *Math. Ann.* **294** (1992), 755–764.

[72] A. PRESSLEY & G. SEGAL, *Loop groups*, Oxford University Press (1986).

[73] T. RATIU & P. VAN MOERBEKE, The Lagrange rigid body motion, *Ann. Institut Fourier* **33** (1982), 211–234.

[74] A. G. REIMAN, Integrable Hamiltonian systems connected with graded Lie algebras, *J. Soviet Math.* **19** (1982), 1507–1545.

[75] A. G. REYMAN & M. A. SEMENOV-TIAN-SHANSKI, Reduction of Hamiltonian systems, affine Lie algebras and Lax equations I *and* II, *Invent. Math.* **54** (1979), 81–100 *and* **63** (1981), 423–432.

[76] A. G. REYMAN & M. A. SEMENOV-TIAN-SHANSKI, A new integrable case of the motion of the 4-dimensional rigid body, *Commun. Math. Phys.* **105** (1986), 461–472.

[77] A. G. REYMAN & M. A. SEMENOV-TIAN-SHANSKI, Group theoretical methods in the theory of finite dimensional Integrable systems, in *Dynamical systems VII, Encyclopaedia of Math. Sci., Springer* **16** (1994).

[78] E. REYSSAT, *Quelques aspects des surfaces de Riemann*, Progr. Math. **77**, Birkhäuser (1990).

[79] M. A. SEMENOV-TIAN-SHANSKI, What a classical r-matrix is, *Funct. Anal. Appl.* **17** (1983), 259–272.

[80] M. A. SEMENOV-TIAN-SHANSKI, Lectures on R-matrices, Poisson-Lie groups and Integrable systems, *in* [16].

[81] J.-M. SOURIAU, *Structure des systèmes dynamiques*, Dunod, Paris (1969).

[82] P. VANHAECKE, Linearising two-dimensional integrable systems and the construction of action-angle variables, *Math. Z.* **211** (1992), 265–313.

[83] P. VANHAECKE, Integrable systems and their morphisms, *preprint* (1994).

[84] J.-L. VERDIER, Algèbres de Lie, systèmes hamiltoniens, courbes algébriques, in *Séminaire Bourbaki*, Springer (1980).

[85] G. WEICHOLD, Über symmetrische Riemannsche Flächen, *Zeit. J. Math. Phys.* **28** (1883), 321–351.

[86] A. WEIL, Euler and the Jacobians of elliptic curves, in *Arithmetic and Geometry, papers dedicated to Shafarevich*, Progr. Math., Birkhäuser (1983).

[87] E. T. WHITTAKER, *A treatise on the analytical dynamics of particles & rigid bodies*, 4th edition, Cambridge University Press (1927).

[88] J. WILLIAMSON, On an algebraic problem concerning the normal form of linear dynamical systems, *Amer. J. of Math.* **58** (1936), 141–163.

[89] S. L. ZIGLIN, Branching of solutions and nonexistence of first integrals in Hamiltonian mechanics I *and* II, *Funct. Anal. Appl.* **16** (1982), 181–189 *and* **17** (1983) 6–17.

Index

Printed in the United States
By Bookmasters